PRACTICAL
PROJECT
MANAGEMENT

FOR BUILDING AND CONSTRUCTION

ESI International Project Management Series

Series Editor

J. LeRoy Ward, Executive Vice President
ESI International, Arlington, Virginia

Practical Project Management for Building and Construction
Hans Ottosson • 978-1-4398-9655-6

Project Management Concepts, Methods, and Techniques
Claude H. Maley • 978-1-4665-0288-8

PgMP® Exam: Practice Test and Study Guide, Third Edition
Ginger Levin, and J. LeRoy Ward
978-1-4665-1362-4

Program Management Complexity: A Competency Model
Ginger Levin, and J. LeRoy Ward
978-1-4398-5111-1

Project Management for Healthcare
David Shirley • 978-1-4398-1953-1

Managing Web Projects
Edward B. Farkas • 978-1-4398-0495-7

Project Management Recipes for Success
Guy L. De Furia • 978-1-4200-7824-4

A Standard for Enterprise Project Management
Michael S. Zambruski • 978-1-4200-7245-7

Determining Project Requirements
Hans Jonasson • 978-1-4200-4502-4

The Complete Project Management Office Handbook, Second Edition
Gerard M. Hill • 978-1-4200-4680-9

Other ESI International Titles Available

from Auerbach Publications, Taylor & Francis Group

PMP® Challenge! Fourth Edition
J. LeRoy Ward and Ginger Levin
978-1-8903-6740-4

PMP® Exam: Practice Test and Study Guide, Seventh Edition
J. LeRoy Ward
978-1-8903-6741-1

The Project Management Drill Book: A Self-Study Guide
Carl L. Pritchard
ISBN: 978-1-8903-6734-3

PRACTICAL PROJECT MANAGEMENT

FOR BUILDING AND CONSTRUCTION

HANS OTTOSSON

CRC Press
Taylor & Francis Group
Boca Raton   London   New York

CRC Press is an imprint of the
Taylor & Francis Group, an **informa** business
AN AUERBACH BOOK

Cover design: Tobias Lindqvist Ottosson

CRC Press
Taylor & Francis Group
6000 Broken Sound Parkway NW, Suite 300
Boca Raton, FL 33487-2742

© 2013 by Taylor & Francis Group, LLC
CRC Press is an imprint of Taylor & Francis Group, an Informa business

No claim to original U.S. Government works

Printed in the United States of America on acid-free paper
Version Date: 20120522

International Standard Book Number: 978-1-4398-9655-6 (Hardback)

Library of Congress Cataloging-in-Publication Data

Ottosson, Hans.
    Practical project management for building and construction / Hans Ottosson.
        p. cm. -- (ESI International project management series ; 11)
    ISBN 978-1-4398-9655-6 (hardback)
    1. Construction projects--Management. 2. Project management. I. Title.

HD9715.A2O88 2012
624.068'4--dc23   2012004660

Visit the Taylor & Francis Web site at
http://www.taylorandfrancis.com

and the CRC Press Web site at
http://www.crcpress.com

Contents

The Author

Hans Ottosson has extensive experience in project management. It began in 1972 with a post as assistant project manager for a new brewery in Sweden. In the spring of 1998 he was persuaded to become project manager for the railway contract above and below the Öresund, between Denmark and Sweden. The project was 7 months behind schedule, but was successfully completed by July 1, 2000.

Hans, who holds an MSc, started his professional career in 1969 as a structural engineer and a teacher of bridge design at Chalmers Technical University in Gothenburg, Sweden. He has worked as project manager for projects in the United States, Saudi Arabia, and Scandinavia. In 1977 he started his project management company, Projsam Quality AB, and has had clients such as Volvo, AstraZeneca, Avesta Sheffield, Skandia, Swedish Transport Administration, and a number of national and local companies.

He has worked as a client representative, structural designer, and contractor and been the project manager for various projects such as construction of production facilities, housing, medical research facilities, and schools. He has organized public conferences and reorganized companies in the IT sector; he has been responsible for international marketing projects.

Hans has developed project management procedures over many years. He was awarded a "Gold Palm" at the PMI conference in 2002 in Cannes for best "tools and techniques." Hans teaches at Chalmers Technical University and for ESI in Stockholm.

1

Introduction

To pursue projects takes more than willpower and a great deal of ambition. You also need knowledge and the right information at the right moment.

The construction industry has been doing projects since Pharaoh Zozer's time, about 3000 BC. Imotep is, as far as I know, the first documented project manager. He was also an architect and the builder of the first pyramid, a stepped pyramid. With thousands of years of accumulated knowledge, you must wonder if we builders really have anything new to learn.

In the mid-1900s space flights, technologically advanced weapons systems, and the computer industry began to run large and complex projects. New tools and techniques to succeed in complex projects were developed, primarily by NASA, the US military, and the IT industry.

Clients began to require some form of knowledge for project managers of complex projects. Needs and requirements for certification of project managers began to emerge. It is as obvious that a project manager should have a basic knowledge of how to run a $3 million project as it is that someone driving a motorcycle have a driving license.

Project management is a profession that requires specific knowledge and skills. Local certification has existed for some time, but has not been widely practiced. Today, there are two major international certification agencies that certify project managers: the Project Management Institute (PMI) and the International Project Management Association (IPMA).

At the end of the 1960s, the PMI compiled tools and processes so that it would be easier to manage various projects. The information was gathered in the PMBOK's® "A Guide to the Project Management Body of Knowledge." This publication, which has been revised several times, is the knowledge base for PMI certification. PMI has certified project management professionals (PMPs) since the certification started in 1984. This certification is limited over time and the PMP must demonstrate that he or she has continued to work with projects in order to retain certification. In 2012, PMI had more than half a million members and credential holders in more than 185 countries.

Another organization that certifies project managers is IPMA. Its certification is done at four different levels, depending on skills and experience. This certification is also time limited. By the end of 2010, there were more than 130,000 IPMA-certified professionals worldwide.

The old dominant team leader with his boots and his helmet, instead of cowboy hat, was very much a John Wayne type. When he saw a problem, he made a swift arc, made a decision, and shot the problem dead. If this in turn created new problems, he shot them down as well. Too often, this resulted in delays, cost, and quality problems.

The new generation of project members with their computer knowledge and easy access to a wealth of information will not accept work in this way. Although there is continuous development of a large number of computer programs as project aids, knowledge of project management tools is often missing. This means that the tools are not used optimally. The new IT tools have allowed employees to become fantastic equilibrists at the keyboard; however, without knowledge of building materials, building physics, design, and management, they are easy prey for marketers of new materials. We must therefore have risk analysis and quality control to help us design and build the right things.

This book describes the processes and knowledge areas that are necessary in order to run successful projects. It also includes basic information about various tools and techniques such as work breakdown structure, earned value, and how to use a network diagram with its gaps and critical paths.

The examples in Chapter 6, "Knowledge Areas," are from industrial construction projects. The knowledge areas, however, are the same for any type of project. The material can therefore easily be applied to other construction projects.

The processes, tools, and methods presented in this book are general and should be used by clients, project managers, team leaders, and designers in their control of work. Large parts of the book describe general project management. The book can also be used by project managers outside the building and construction industry.

Remember to suit the planning work to the size and complexity of the project. If you know **WHAT** to do, **HOW** to do it, **WHEN** to do it, and **WHO** should do it, you are in control of your project. Then execute with your brain, heart, and guts.

Hans Ottosson

2

General Information on Projects in the Building, Construction, and Installation Industries

In the building and construction industry, there are many stakeholders: real estate developers, architects, engineers, contractors, subcontractors, etc. There are also many established types of organizations, contracts, and remuneration. Building/construction is often a local business, which means that names, procedures, etc., have developed differently in different countries and areas. This chapter tries to sort out many of the expressions and procedures and give them names to be used in this book; they may vary from those with which the reader is familiar.

2.1 PROJECT ROLES

When I talk about project managers, subproject managers, contractors, consultants, customers, and clients, I have attempted to facilitate the understanding of the text by using the following terms (organizational charts are shown in Figures 2.1–2.3):

- **The owner** is the client organization.
- **Users** and **end-users** are those who use the product (e.g., tenants, researchers, drivers, train companies, travelers, athletes, and spectators).
- **The project owner** is the person in the client organization who has received the board's mandate to implement an investment. The person is sometimes called the "sponsor," "project sponsor," or "authorizing sponsor." This can be the president, the plant manager, or

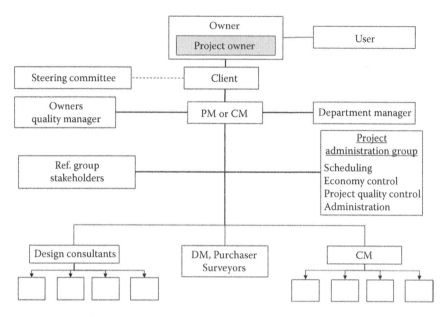

FIGURE 2.1

Project owner organization (project roles; see Section 2.1).

another person designated to implement the project. Sometimes the project owner is replaced with a program governance board.

- **The client or customer** is the owner or ordering organization's representative who orders work and services. This may be the plant manager or another person designated to implement the project.
- **The project manager (PM)** is the client's project manager. Sometimes, the project owner, client, and the project manager—or the client and project manager—are the same person.
- **The task leaders (TLs)** are the architect's and consultant's project managers (one of the project manager's subproject leaders) for the project.
- **The site manager (SM)** is the client's representative on the building site, a subproject leader to the PM. In many areas, he or she is called a team leader, but I have chosen SM so that this will not be confused with the design task leader.
- **The contracts manager (CM)** is the main contractor's project manager (one of the project manager's subproject leaders) for the project. In many areas, he or she is called a team leader but I have chosen the CM so that this will not be confused with the design task leader or the SM.

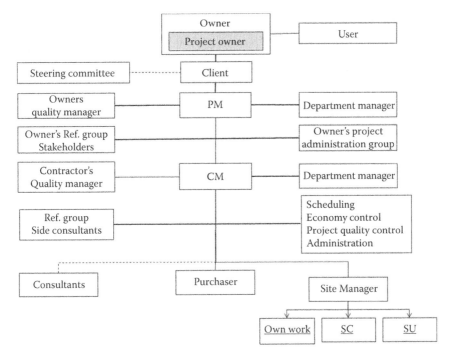

FIGURE 2.2
Main contractor's organization. Note that the project manager is the contractor's customer. The dotted line to the consultants is valid for FuP/DC contracts (SC = subcontractor and SU = supplier).

- **The installations managers (IMs)** are the installation contractors' project managers (the project manager's subproject managers and/or the main contractor's submanagers).
- **The designs manager (DM)** is the subproject leader responsible for all the design work. Responsibility for the technical solutions is usually left with the respective design consultant. In smaller projects, the PM often has the role of DM.
- **Project members** are those working on the project. Members can be the client, users, project planners, contractors, installers, and suppliers.
- **The steering committee** is a management team assembled to support the project owner and project manager in all matters regarding the project. The project owner is often the president of this committee.
- **The reference group** is an advisory group of interested parties and experts and often represents future users of the project result.

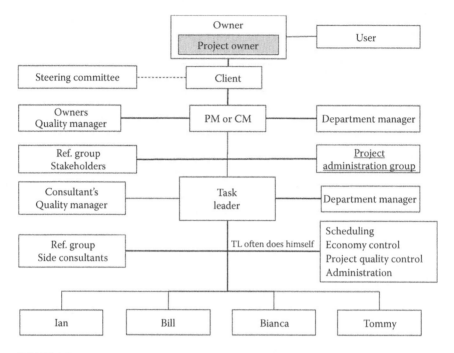

FIGURE 2.3

The task leader's organization. Project roles: the purchaser of the design is either the project manager (design–bid–build contracts) or the contracts manager (at FuP/DC contracting). (See Section 2.5.2.) Types of contract: often, the task leader (TL) is doing many of the project administration tasks in the design office.

2.2 STAKEHOLDERS AND ORGANIZATIONS

Project stakeholders. Everyone involved in a project is called a project stakeholder. In the building/construction industry, there are

- Property owners
- Developers, owners, project owners
- Steering committee members
- Clients or customers
- Users and end-users
- Operations and maintenance staff
- Project members:
  - Project managers, schedulers, calculators, design managers, purchasers, construction managers, inspectors, surveyors

- Planners; architects; structural, electrical, and HVAC engineers; electricians; land developers; etc.
  - Contractors, builders, installers
  - Suppliers, carriers
  - Consultants in various specialized areas
- People in areas such as finance, purchasing, and production departments
- Authorities involved in the procurement of a third-party interest
- The public affected by the project
- Project members' families
- Trade unions, employer organizations, and trade organizations

Although all stakeholders are affected in one way or another in our projects, all stakeholders are not equally important. The most important group (core stakeholders) will vary from project to project but usually include:

- Owner/project owner
- Client or customer users of the product
- Project manager, main contractor, installation managers, and task leaders
- Project members (including consultant's and contractor's employees)
- Steering committee members
- Authorities

Traditionally, the client focus is very strong. Occasionally, a PM even forgets the end-users (customers' customers) when working on a project. Therefore, when houses are built, it is necessary to focus on the entire product user chain, from the global to the local.

Other key stakeholders are

- Partners (client partners, the consortium)
- Secondary suppliers to the project (e.g., subcontractor's suppliers' suppliers)
- The public

Among the secondary stakeholders are

- Competitors to clients, designers, contractors, suppliers, etc.

- Associations (e.g., political, trade, and advocacy organizations)
- Competing projects
- Media (information; negative or positive publicity)

Stakeholders and their needs must be identified and evaluated. Sometimes stakeholder needs are not identified or understood. This can result in a project closing down or long delays. Stakeholders often have different needs. Client, designer, and contractor expectations are not always consistent.

Often there are conflicts between different stakeholders. Sometimes it is the PM's greatest challenge to settle conflicting stakeholder demands. Conflicts between stakeholders need to be resolved early so as not to cause delays and increase costs. Stakeholders must understand and accept the compromises made. The PM must be aware that the outcome of a project is assessed as the product of acceptance and quality ($R = A \times Q$). There must be funds in the project budget for the management of stakeholders.

The project manager must not be indifferent to making decisions, but should be responsive and able to adapt. These are two main ingredients in the work with stakeholders. The PM must not be afraid or hesitate to make decisions. He must be receptive to other ideas and be able to adapt to new situations.

2.2.1 Examples of Project Organizations

The organization chart in Figure 2.1 shows a **design–bid–construct (DBC) contract** (see Section 2.5.2 and Figure 2.7 later). Note that the owner's quality manager in Figure 2.1 is to the left of the project manager and the quality manager of the project is in the project administration group to the right. The dotted line to the steering committee indicates that not all projects have a steering committee.

In a functional/performance (FuP) or design/construct (DC) contract (see Section 2.5.2), the main contracts manager is the designers' "customer." Accordingly, the designers must also take into account the contractor's business plan and quality plan. The organizational chart for a main contractor's organization in a functional contract is shown in Figure 2.2.

In all organizations, the PM, the CM, and the TL must manage their projects and contracts in terms of time, economy, quality, risk, and other areas of knowledge, which are described later in Chapter 6.

2.2.1.1 Matrix Organization

Within a company, such as in a consulting company, there is often a matrix organization with line managers responsible for different areas of expertise and task leaders responsible for the assignments (see Figure 2.4).

The line managers for the various specialist areas are responsible for the employees' having the requisite skills. The responsibility for the technical solutions in each department is on the line manager unless the project manager has taken over—in writing (e.g., in the minutes of a meeting)— the responsibility for a certain design. The latter is relatively uncommon. The line manager should also coordinate and allocate resources between the various projects. It must be clarified if the TL or line manager is responsible for the recruitment of outside assistance if certain skills or capabilities do not exist.

2.2.1.2 Steering Committee

The steering committee's purpose is to facilitate the project, just as the project manager will facilitate for the project staff. The steering committee

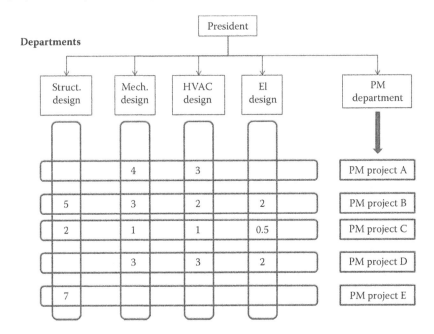

FIGURE 2.4

Matrix organization showing shared responsibility within a design engineering company. The figures show the number of engineers who work for different projects.

considers changes to the project objectives and constraints and facilitates project integration within the organization.

The steering committee often has the project owner (authorizing sponsor) as president and line managers or resource members (reinforcing sponsors) that will prioritize resources to the project. For the project, it is important to document the steering committee's decisions quickly. If the project owner does not attend the committee meetings and there are changes to the project objectives and constraints, these changes must be approved by the project owner. The project manager takes a chance if changes are implemented before the change decisions are documented **in writing.** The PM communicates primarily with the committee president and should stay away from power struggles, which are not uncommon in this type of committee.

The main mission for the steering committee, the PM, the CM, and the TLs is to help with decisions, support project members, and otherwise facilitate the project work. It is always the manager's job, at all levels, to "rake the circus ring" for those who must carry out the work.

The main task for the steering committee is often to help the project manager to obtain resources (personnel and equipment) from the different line managers.

2.3 INTEREST GROUPS

2.3.1 Interest Groups in the Building and Construction Industry

In order to facilitate construction work, a number of codes, procurement, contracting, and remuneration tools have been developed and have evolved over the years. Many of these were developed jointly by trade organizations for builders, contractors, engineers, architects, etc. Some trade organizations are

ABC	Associated Builders and Contractors (United States)
AGCA	Associated General Contractors of America (United States)
AIA	American Institute of Architects (United States)
ASCE	American Society of Civil Engineers (United States)
ASHRAE	American Society of Heating, Refrigerating and Air-Conditioning Engineers (United States)

BBK	Byggandets kontraktskommité (Sweden)
BOCA	Building Officials Code Administrators International (United States)
CMAA	Construction Management Association of America (United States)
ECA	The Electrical Contractors Association (United States)
FIDIC	International Federation of Consulting Engineers (Fédération Internationale des Ingénieurs-Conseils) (international)
IAPMO	International Association of Plumbing and Mechanical Officials (United States)
NCMA	National Contract Management Association (United States)
RIBA	Royal Institute of British Architects (UK)
	Western Fire Chiefs Association (United States)

More information about other associations can be found on the Internet.

2.4 WAYS TO IMPLEMENT CONTRACTS

A developer can build in-house or hire consultants and contractors. Normally, a developer does not have adequate resources and skills to produce drawings and specifications for projects. This means that the developer or contractor (subject to contract terms) needs to purchase design services.

The contractor also needs to purchase materials and subcontracts—for example, sheet metal work, electrical installations, and plumbing. How these purchases will be executed depends on what the market looks like. Does the procurement have a large or small impact on project economy and time? Are there many suppliers or only one? For more on purchasing strategies, see Section 3.5.2 in Chapter 3.

Contract implementation consists of four parts (see Figure 2.5):

- Type of procurement procedure
- Deliverable or type of contract (e.g., DCM, TK, DC, DBC, EPC; see Section 2.5.2)
- Type of remuneration (e.g., fixed price, cost plus; see Section 2.6)
- Type of cooperation

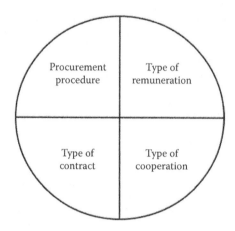

FIGURE 2.5
Parts of procurement.

The following questions must be asked:

WHAT shall I buy?—Designs and/or constructions, installations, functions, products, and/or maintenance (for type of contract, see Section 2.5)

HOW shall I buy?—Ask for bids, buy over the counter, or only negotiate for an agreement? (for contract procedure, see Section 2.4.1)

WHAT type of remuneration do I want?—Payment as a fixed price/lump sum or as cost plus or reimbursement based on unit prices (for type of remuneration, see Section 2.6)

HOW should the seller and buyer cooperate?—How to get a win–win situation for buyers and sellers? Help with financing? Work together to develop new technologies while the responsibility for function and maintenance and the apportionment of the profit is fair? (for type of cooperation, see Section 2.7.)

2.4.1 Types of Procurement

What does the market look like? What contract procedure gives the best—not cheapest—project solution? This depends on the market situation and desire for long-term relationships.

US authorities impose specific requirements on state and federal agencies, counties, etc., when they are buying goods and services. Similar requirements exist in Europe. These regulations limit the public purchaser's discretion over how a private buyer can act.

The procurement should be competitive. If this is not possible because there is only one supplier (licensed owner or acceptable supplier), an agreement must be negotiated, which may require considerable time and skill from the purchaser.

The client must act ethically in all negotiations. Many companies have special codes of ethics that helps employees when awkward situations arise. To act unethical means rogue negotiations and failed projects in the long run.

Providers should fulfill their obligations under applicable laws and regulations including safety, health, environment, unlawful discrimination, and often labor union relations. The following procedures have been identified:

Open procurement. All suppliers who wish and possess the technical capability to fulfill the requirement may submit bids.

Selective bidding. The client decides who may bid. These can be selected through a prequalification process.

Negotiated procurement. The client negotiates with a selected number of vendors in the competitive range. Discussions are held with these vendors to identify the strengths and weaknesses of their proposals and finally to agree on type of remuneration and other contract issues. Negotiated procurement is also used when there is only one acceptable vendor.

Direct award. The customer buys directly without written bids. Laws and regulations normally limit state agencies' use of this option to purchase.

Electronic bidding, online bidding, and reverse auction online. Contractors may submit bids on the Web. These bids are known to all at the same moment that the offer is made. Other parties then have limited time to submit competitive bids. This is a relatively new type of bidding for the construction industry. The reverse auction online has lost traction over the years.

Contracting in one or two steps. In FuP and DC contracting, the procedure for bids can be in one or two steps:

When the one-step procedure is used, technical solutions and price are requested at the same time. However, the solutions are evaluated without seeing the price. When there is a short list with technically acceptable bidders, their prices and other attributes are evaluated.

When using two steps in contracting, the purchaser may, as a first step, call for proposals of technical solutions (request for proposal; RFP). The bidders that are technically acceptable will then price their solution in a second step. When contracting in two steps, it is important to maintain competition in the second step. Thus, in the first step, a winner must not be announced (e.g., in an architectural contest); rather, the acceptable proposals should simply be noted. These bidders can then move on to the pricing process.

The **best** solution for the client is not always the cheapest or most technologically advanced. The best solution in all respects is the one that should be purchased.

Tender process. It is important that the PM take part in the budget process and that the TLs, CM, IMs, and senior technicians take an active part in the tender process and negotiations. Contract knowledge is the key element to identifying quickly what should be done.

2.5 TYPES OF CONTRACT AGREEMENTS

2.5.1 General

A contract agreement is a mutually binding agreement that obligates the seller to provide the specified product or service and obligates the buyer to pay for it. In order to form an enforceable contract, the parties must be competent and legal entities (companies) or private legal persons. Individuals representing companies are called agents. An agreement creates an obligation to do or not to do a particular thing.

Agreements can be written or verbal. If a dispute arises about the verbal agreement, then a burden of proof is on both parties. Written agreements should be created to confirm verbal agreements.

An agreement in the construction industry must include information about the parties' specified product, performance, and commitments; the conditions for execution; and the compensation. Governments in most countries have enacted a number of general laws about contracts that control the parties' lines of conduct.

In many countries, there is a lack of specific legal rules for the building and construction industry. Therefore, trade organizations (developers, planners, consultants, contractors, installers, and suppliers) have established specific rules for the industry. For example, the International

Federation of Consulting Engineers (FIDIC) has published forms for different types of contracts:

- "The White Book Guide with Other Notes on Documents for Consultancy Agreements"
- "Client/Consultant Model Services Agreement"
- "Conditions of Contract for EPC/Turnkey Projects"
- "Conditions of Contract for Construction"
- "Conditions of Contract for Plant and Design Build"
- "Short Form of Contract. Agreement, General Conditions, Rules for Adjudication, Notes for Guidance"
- "Supplement to Third Edition 1987 of Conditions of Contract for Electrical and Mechanical Works"

Business contracts can also include financing. For information on this, see Section 2.7.

A number of contract types are based on bidding documents and deliverables without financing:

- Concept contracts
- Functional/performance (FuP) contracts
- Design/construct contracts
- Design–bid–construct contracts; also called engineering, procurement, and construct (EPC) contracts
- Management organization contracts
- Subcontracts (SCs)
- Side or joint contracts (project work not included in the main contractor's scope)
- Suppliers
- Services

2.5.2 Types of Contracts

2.5.2.1 Concept Contracts

When looking for ideas about what to do with a building site or an entrance hall, a concept contract may be used. This type of contract can be an artistic adornment in a building, for example. It is hard to find contract support documents from our trade industry, but a contract may be based on documents for design, construct, and maintain (DCM)/design and construct (DC) or turnkey (TK) contracts.

2.5.2.2 Functional/Performance Contracts

Design, construct, and maintain contracts. The client describes a function. The contractor is responsible for the design, production, and an extended functional responsibility over a contract period during normal production. Often the contractor maintains the project or part of the project during this time. The contact work will not be completely taken over until all functional requirements have been tested and met during a period of normal production. The functional contractor is responsible for hiring a design team. He or she will also contract with suppliers and subcontractors such as sheet metal workers, plumbers, and electricians. This type of contract is assumed to provide the contractor more freedom to develop new technologies without increase in the customer's risk.

The request for proposal should contain not only a functional description, but also rules for functional controls over the contract period. Functional/performance contracting involves extended contractor responsibility compared to a DC contract.

The project may have client work and client-purchased side contracts (CPSCs), such as machinery. It is the client's responsibility to perform and coordinate this work with the main contractor's work.

Sometimes performance contracting allows a facility to complete improvements within an existing budget by financing them with money saved through reduced utility expenditures. This type of performance contract needs no up-front investments; instead, the project is financed through guaranteed annual savings.

Turnkey contracts. A turnkey contract is a DC/DCM contract (see Figure 2.6) that normally includes a more enlarged delivery; for example, it can include furniture and office equipment. The users simply get the "key," walk in, and start to work. FIDIC provides contract support through "Conditions of Contract for EPC/Turnkey Projects."

Design and construct contracts. The contractor is responsible for the design and production. The DC contractor is responsible for the design team's work to the client. He or she will also contract with suppliers and subcontractors such as sheet metal workers, plumbers, and electricians. The contract work will be taken over by the client when

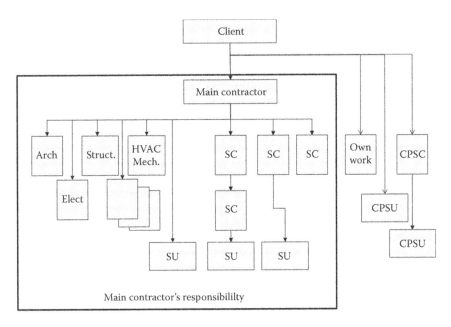

FIGURE 2.6
Organization of FuP, DC/DCM, and TK projects. Note the responsibility line between the CPSCs and the main contractor. This requires coordination.

the project work has been tested after completion and the client issued a "taking-over certificate." After the takeover, there will be no functional responsibility except for limited traditional defects liability. FIDIC provides contract support through "Conditions of Contract for Plant and Design-Build."

Guided DC (GDC) contract—sometimes called a develop, design, construct (DDC) contract. In GDC contracts, the client has designed the product to a certain level and then purchases it as a DC, DCM, or TK contract. With this method, the client can decide the building and installation systems and sometimes even more details. By purchasing with a DC, DCM, or TK contract, the client may transfer the risk of inaccuracies in the consultant's design work to the main contractor (depending on the legal agreement text). On the other hand, the client may limit the contractor's use of production experience, special equipment, and special skills.

Design, bid, construct; engineering, procurement, construction; and design-to-build contracts. With the help of consultants, the client obtains drawings, specifications, and other inquiry documents. The contractor performs as designed in accordance with the contract. Under this contract type are the following construction types of contracts:

General construction (GC) contract. The general contractor engages subcontractors and suppliers and together they perform what the client (often with the help of consultants) has designed and described. FIDIC provides contract support through "Conditions of Contract for Construction" (see Figure 2.7).

Coordinated general construction (CGC) contract. For a fee, the main contractor (normally the building contractor or general contractor) coordinates the client work and CPSCs. The client chooses the coordinated general construction model to be able to influence the purchase of subcontractors or installers who will work with the project. Sometimes the client has procured long-lead-time items early in order not to lose time and then transfers this contract to the main contractor. The main or general contractor's undertaking includes scope, risks, coordination of the CPSC work, and his or her own work.

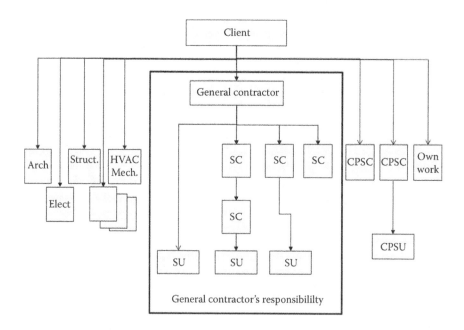

FIGURE 2.7

Organization for general construction. Note the responsibility line between the general contractor and the design team and the CPSCs. This requires coordination.

2.5.2.3 *Management Organization Contracts*

Divided contracts. The client buys a number of contracts that he or she will coordinate or, for a fee and limited liability, transfer to a project management organization. The different contracts purchased by the client can be FuP, DC, or DBC or a supplier (see Figure 2.8).

  Subcontracts. This is work purchased by a contractor. The contractor is responsible to the client for the scope, time, costs, risks, and coordination. A subcontract can be an FuP, DC, or DBC contract or a supplier.

  Services contract. Services contracts cover a broad category of items. This type of contract provides for the sustainment of performance at the site. This can be site service, call center, or other project-oriented business activities.

2.5.3 Consultant Agreements

The client often lacks design skills and must therefore purchase these services from consultants. The requirements for documents are different in DC and DBC contracts. Bidding documents will become the legal documents between client and contractor and must contain technical and legal

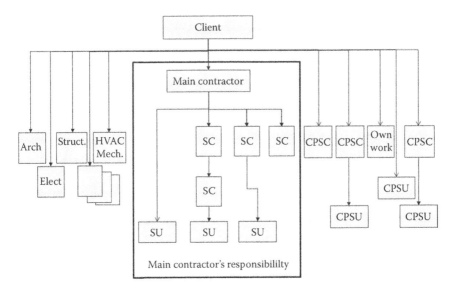

FIGURE 2.8
Organization with divided contracts. Note the responsibility lines between the main contractor and the design team as well as the client works and all the CPSCs. This requires qualified coordination.

sections. In a DC contract, this type of document can be kept down. It may be needed when dealing with suppliers and subcontractors. When one builds with one's own personnel, the team leader can give verbal instructions and therefore not as many detailed drawings and specifications are needed. Consultant's agreements must be adapted to the type of contract. FIDIC provides contract support through the "Client/Consultant Model Services Agreement."

The consultant contract is sometimes divided into phases, based on different design stages (see Section 6.12.3 in Chapter 6). The following documents (see Section 5.4 in Chapter 5) will be produced:

- Project requirement documents
- Feasibility study
  - Functional and building program
  - Design outline specification
  - Project management plan (including time schedule and governing documents for economy, risk, procurement, organization, etc.; see Sections 3.5.1, 4.2, and 5.3 in Chapters 3, 4, and 5, respectively)
- System and basic design drawings and documents
- Inquiry and bid documents
- Construction and building documents
- As-built documents

Clients will sometimes launch a design competition for architects. Trade organizations for architects have agreed on fair competition terms in many countries and states. Another way to get competition between different architects or consulting groups is to pay some compensation to a chosen number of participants for their proposals—so-called parallel studies.

2.6 TYPES OF REMUNERATION

Since a project is unique and not a production series where the costs of long runs can be studied and then the product priced, uncertainty is built into pricing. The type of remuneration is a way to distribute the risk of economic uncertainty between the contract partners (see Figure 2.9). There are various types of remuneration for building contracts.

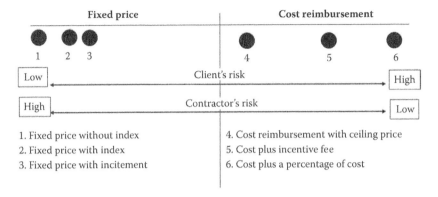

FIGURE 2.9
Client and contractor risks for different types of remuneration.

Fixed price or firm-fixed price means that the contractor will be reimbursed with a fixed sum for execution of the work. This amount includes the cost of administration, interest rates, risk, and profit. Fixed price can also be combined with an incentive.

Cost reimbursement means that the contractor gets paid for verified expenses. In addition, the contractor receives a fee to cover administrative costs, interest, and profit. Sometimes, for the sake of simplicity, the remuneration is based on agreed prices per unit, called **time and material.** There are many forms of cost reimbursement: cost plus award fee, cost plus incentive fee, costs plus fixed fee, and cost plus a percentage of cost.

Quantity contracts are based on measured quantities from drawings, sketches, and specifications prepared by designers, principally architects and engineers. The quantities of work "taken off" from drawings are used to prepare a bill of quantities (BoQ) or bill of material (BoM). The contractor will then price this quantity document in a competitive bid and be paid according to a measure undertaken on site and applied to each specific work item.

In Scandinavia, this is sometimes used for earthworks, landscaping, and rock excavations. In the UK, Australia, and New Zealand, this method is used for all kinds of building and construction. These remuneration forms are often supplemented to spread risks (uncertainties) and development opportunities and to simplify the presentation of receipted costs.

For projects that run for a long period, the fixed price or fixed price per unit is combined with an adjustment clause for price (index). For materials and machinery purchased abroad, the exchange rate can change during the project. In this case, an exchange rate must be decided when the contract is written and then adjusted when the machinery is purchased.

In order to reduce customer risk, the cost reimbursement can have a ceiling price—that is, the maximum amount that the buyer will pay. If the client and the contractor have acknowledged that neither of them knows (with any certainty) how much it is going to cost to perform the work, they can decide to agree on a target cost and a formula for sharing the overrun or underrun that is almost certain to occur.

The parties may establish an incentive contract. In this type of contract the client and contractor determine target price. If the price is being undercut, the client and contractor share (e.g., 30/70) the difference—fixed-price-incentive, firm (FPIF) or cost-plus-incentive fee (CPIF).

2.7 TYPES OF COOPERATION

Sometimes participants in the construction industry choose various types of cooperation. Owners and contractors or just contractors form consortia to reduce risks or complement each other's skills and resources. The consultants form project teams of specialist consultants to offer clients a group whose members are accustomed to working together.

Some types of cooperation are chosen to solve issues of funding or promoting new technologies and construction methods without increasing the customer's risk. Many types of cooperation have been introduced on infrastructure projects and, as the experience has increased, it has been utilized in other major projects.

Common forms of cooperation include

BOT/BOOT Build–operate–transfer/build–own–operate–transfer. BOT or BOOT is a project form where a private entity receives a concession from the private or public sector to finance, design, construct, and operate a facility stated in the concession contract. When compensation through revenue has covered the construction, operating costs, and agreed profit, the ownership of the facility and its operations is transferred to the contracted organization.

PPP, P3 Public–private partnership. PPP is a joint project where the public sector accounts for all or part of the investment

	and the private company is responsible for the operation, or a private company accounts for the investment and the public is responsible for all or parts of the operation.
PPC	Public–private cooperation. PPC is a collaboration between public and private sectors. The cooperation is mainly project management and procurement. In Sweden, it has been used for infrastructure projects. In 2010 it was introduced at the New Karolinska Solna University Hospital in Stockholm.
Partnering	The goal of partnering is to focus on common goals and reduce the optimization of specific partner interests. The contractors take early part in the design stage to facilitate the company's production skills and production methods. The parties sign the agreement with the understanding that they will share proportionally the losses and profits according to the provisions and conditions that they have mutually assented would govern their business relationship.

More recently, new forms of cooperation have been introduced in Scandinavia by various contractors, including

- "Coopetition"
- Lean construction

I regard construction management as a project type (see Section 2.8), not a type of cooperation.

2.7.1 Cooperation with Building Information Management

Today, large projects are designed in three dimensions (3D). Originally, 3D was a visualization tool, but it has developed into a tool with information on quantities, factory production, schedules, production data, energy balances, and management information (4D and 5D). Building information management (BIM) is a structure and approach that not only shows what the project will look like, hopefully, but also reduces the risk of coordination misses between architectural, structural, mechanical, and electrical drawings. It can supply production drawings (e.g., for the steel production, lists of materials, lengths of pipes and ventilation ducts, etc.). BIM can also provide information for production management, scheduling, and management information.

Sometimes BIM is translated as "building information model." In the international literature, it is often translated as building information management, which is better as it shows that all participants work with the same information in a common database.

BIM is not the same as 3D, although the visualization part is included. The BIM computer models are different as they are based on databases and not a set software program or a shared server.

Working with BIM provides project advantages. In particular, it gives increased and easier information to all stakeholders. Coordination is very important in tight spaces (such as above suspended ceilings) and in power plants with a need for simple operation and maintenance. Drawings done by different designers and suppliers are not easy to understand if they are not compounded in one model. With BIM databases, many problems on the building site can be avoided.

As project managers and task leaders, we must be vigilant. It is important that designers not only have IT skills but also have good knowledge of building physics, material properties, construction techniques, etc.

2.8 TYPES OF PROJECT MANAGEMENT

If the customer does not have its own project management organization, it must supplement or rent such a team. This is done according to two models.

Professional independent project management organizations and quality surveyors. PIPMOs and QSs can include project managers, planners, calculators, buyers, construction managers, inspectors, etc. This group with special expertise in project management is composed of independent, sole proprietors or provided by major consulting firms. QS is common in the UK, Australia, and New Zealand.

The independent role on the team comprising client, architect, engineer, and contractor has given the PIPMO or QS a reputation and appreciation for fairness. This, combined with expertise in drafting and interpretation of contract documents, enables him or her to settle issues, value the work fairly and regularly, project final costs, avoid disputes, and ensure the effective progress of a project.

Construction costs are controlled by accurate measurement of the work required on a regular basis; application of expert knowledge of costs and prices of work, labor, and materials; and understanding of the implications of design decisions at an early stage to ensure that good value is obtained for the money to be expended.

Construction management. Construction (project) management (CM) is the overall planning, coordination, and control of a project from inception to completion. It is aimed at meeting a client's requirements in order to produce a functionally and financially viable project that will be completed on time within an authorized cost and to required quality standards. Project management is the process by which a project is brought to a successful conclusion:

- Independent CM company (can be considered as a PIPMO)
- A contractor's project management organization, which is available to the client

2.9 REGULATORY REQUIREMENTS FOR THE CONSTRUCTION INDUSTRY

In order to protect citizens, third parties, and other interests related to construction projects, a number of laws, ordinances, and regulations must be followed. Building codes provide minimum standards for the protection of life, limb, property, and environment and for the safety and welfare of the consumer, general public, and owners and occupants of residential buildings regulated by this code.

With the Occupational Safety and Health Act of 1970, Congress created the Occupational Safety and Health Administration (OSHA) to assure safe and healthful working conditions for working men and women by setting and enforcing standards and by providing training, outreach, education, and assistance.

In the United States, there are a number of codes for different states and municipalities. The International Code Council (ICC) has completed an international codes series that includes

- Building code
- Residential code

- Fire code
- Plumbing code
- Mechanical code
- Fuel gas code
- Energy conservation code
- Performance code
- Zoning code
- Green construction code

Other codes are

- BOCA national building code (BOCA/NBC)
- National electrical code
- NFPA 5000 building code
- Standard building code (SBC)
- Uniform building code (UBC)
- Good laboratory practice (GLP)
- Good manufacturing practice (GMP)

CODES ARE LOCAL BUSINESS. ALWAYS CHECK WHAT IS VALID IN A PARTICULAR BUILDING AREA!

Before a project can be started, permits must be obtained. Keep in mind that the authorities' handling times can often be lengthy.

3

Projects

3.1 PROJECT BASICS

3.1.1 Project Management

A number of guidelines have been developed to support a project manager's work. These guidelines briefly describe how projects should be controlled. One of the guidelines is PMBOK®, published by PMI. Project managers work with a project management process and knowledge areas such as time, cost, uncertainties, procurement, etc.

3.1.2 What Is a Project?

In the industry, the word *project* is used for different activities. Sometimes a line of work is improperly described as a project. Line work is performed according to routine and instructions. One must also distinguish between **project** and **product.**

- **Routine:** specified way to perform an activity
- **Project:** a **temporary** effort to create a **unique** product, service, or result within a specific **cost** projection
- **Product:** the result (routine, new organization, building, equipment) that the project delivers

To create a new model car can be a project, but to produce one of many similar cars is not a project. The production is controlled by procedures and processes (ISO 9000, etc.). Surely car building has a start and an end, but the production of cars is not unique.

A project is not a copy or repetition of what has been done before. Parts of the project or a project process may be similar to those of previous projects, but because the project is unique, project management processes and knowledge areas should be used.

Project	Product
Build the Öresund bridge and tunnel between Sweden and Denmark	Öresund bridge and tunnel between Sweden and Denmark
Design and produce an ATM (automated teller machine) prototype	ATM prototype
Develop and install an IP/telephone system for a customer	User IP/telephone system
Prepare basic design documents to be approved by the client	Basic design documents

A contractor's project, such as building an industrial building, is known as an external project. This type of business project, with external revenue, often has a specified task in terms of product scope/performance, time, and cost (e.g., design–bid–construct, DBC, contracts). A design and construct (DC) project includes development, design, and construction but is also limited by time and cost. The development work, which determines the functional and technical solutions, is limited by the functional descriptions and the contract. The contractor's project starts with the bidding or immediately after signing the contract.

These external projects may only be subprojects from the client's point of view. The client's project also includes funding, commissioning machines, training staff, etc. From a client's perspective, many projects are considered internal projects. The funding is internal and the time of start and completion can be adapted to the client's own operations. Typical internal projects are change projects such as change management, premise or equipment changes, increases in production, and organization changes.

3.1.2.1 The Project Triangle and Project Pyramid

The project must be specified in such a way that whether the project requirements have been fulfilled can be measured. The description of the project's limitations, time, budget, and size/performance should be "SMART as H...":

    Specific   The specific goal and limitations should be clearly described.

Measurable	The goal must be measurable. It must be possible to measure whether the target (such as budget, height, surface, air changes, energy use) was met.
Accepted	Users, clients, and suppliers should all accept the goals.
Realistic	The realistic goal should be reasonably attainable. (Reasonable is relative. Think of John F. Kennedy's 1961 statement about "**achieving the goal, before this decade is out, of landing a man on the Moon and returning him safely to the Earth**.")
Time specific	Specific timing of when the project can start and should be completed should be described with the date, week, and month.
"as H"	The goal must be anchored and embraced by the project participants.

Often the targets are not SMARTly described when the project is initiated and must therefore be developed in the early project phases. (See also Sections 4.1.2, 4.1.4, 5.2, and 6.3 in Chapters 4, 5, and 6, respectively.)

The project's limitations (scope/performance, time, and budget) are often described as the sides of a project triangle (see Figure 3.1). A change to any of the sides will affect one or both of the other sides. This is important to consider when an amendment that lowers budget, shortens time, or increases scope/performance is being considered. The change means that there will be effects on the other limiting parameters.

Maintaining a customer can be as important as time, budget, and scope/performance. Meeting the requirements of the former, but destroying

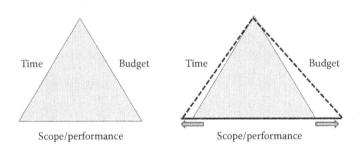

FIGURE 3.1
Project triangle. A change to any of the sides affects one or both of the sides.

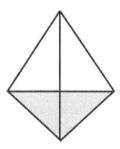

FIGURE 3.2
The project pyramid. The sides represent cost, time, scope/performance, and customer relations. A change to one side will affect one, two, or three sides.

relationships with the client or project owner reduces the chance to get new assignments. Therefore, one more parameter is needed as a project constraint: customer relationship (see Figure 3.2).

With four project-limiting parameters, a pyramid is created. The project pyramid sides describe the project time, budget, scope/performance, and customer relations. To change any of the sides will affect one or more of the other sides in the pyramid. For example, if scope, budget, time, or customer relationship is changed, the pyramid size will be affected.

If the pyramid is inscribed in a sphere that describes the project stakeholders, this will be an interesting picture of the project (see Figure 3.3). If the number of stakeholders and their need for consideration increase, the sphere swells. This will result in a modified form of the pyramid (i.e., there will be an impact on time, cost, size/performance, or customer relationship).

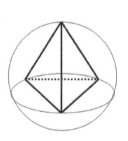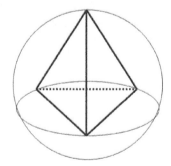

FIGURE 3.3
The project sphere representing stakeholders. A change in the volume of work due to the consideration of stakeholders beyond what is expected will increase the size of the sides of the pyramid and thus affect time, cost, size/performance, customer relations, or more than one of these parameters.

3.1.3 Project Management Process

The project management process has five principal parts: initiation, planning/replanning, implementation, monitoring/acting (control), and completion (see Figure 3.4). These process parts are valid for the entire project, part of the project, a special phase, or a working moment.

- **Initial:** The initiating process answers the questions: **WHAT** should be done? **WHAT** is the scope/performance, budget, and time? **WHO** is responsible for delivering the project? **WHAT** are the authority and responsibilities for the project owner, PM, CM, etc.?
- **Intermediate:** The *planning* process answers the question **HOW** and **WHEN** something will be done and **WHO** should perform the different work packages. Baselines, the reference documents, are created for comparison later when the project is monitored and controlled. In the *implementation*, the job is carried out the way it was planned. Throughout the implementation phase, one *monitors* how the work is progressing and *acts* if there is deviation from what was planned. If the plan is not followed, it is necessary to replan and then implement the new plan.
- **Completion:** In the closure process, the project is approved, filing is done, experiences are documented, and economic, legal, and customer-related activities are finished.

The process is general and describes not only the project but also parts of the project, such as the basic design phase, erecting a superstructure,

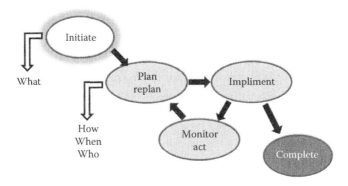

FIGURE 3.4
The project management process is general and describes not only the whole project, but also parts and phases.

installation of machinery, or integrated testing. The process also applies to how knowledge areas are managed, such as

- **WHAT** permission or approval from the authorities is needed? **HOW** is this obtained? **WHEN** is it necessary to apply and **WHO** should do it?
- **WHAT** qualities should be controlled? **HOW** and **WHEN** is this done and by **WHOM?**
- For **WHAT** area should a risk analysis be developed? **HOW** should this be accomplished (method of identification, analysis, etc.)? **WHO** should participate and **WHO** ensures the implementation of planned measures to reduce the threats?

More about the areas of knowledge can be found in Chapter 6.

3.2 PROJECT FLOW

A project flow shows a series of logically related activities that start with an idea or a mission and hopefully end with completion of an approved project. Project flow indicates what will happen from the project initiation until the project is completed. Each phase ends with some form of documentation with an attached approval by the client or project owner. The basic flow, according to PMBOK® in Figure 3.5, shows the chronology of a project. The flow is general, which means that each phase can be divided in the same way. A comparison between PMBOK® and construction project stages and road construction may look like the following (see Figure 3.6):

A client's project is normally broken down into subprojects, such as the designer's or the contractor's projects (see Figure 3.7).

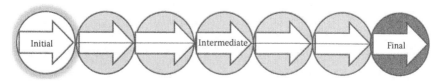

FIGURE 3.5
Project flow according to PMBOK®.

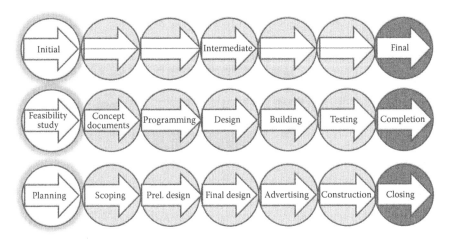

FIGURE 3.6
Comparisons of the project flow in different sectors.

The flow of a normal construction project during the implementation phase is described as follows; there are three different processes (see Figure 3.8):

The design flow

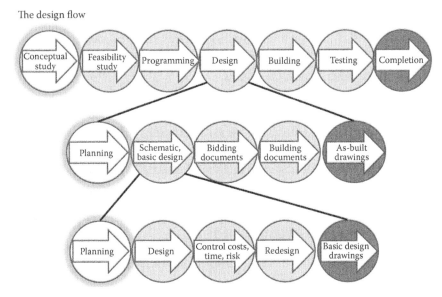

FIGURE 3.7
The design flow broken down into subdesign flows.

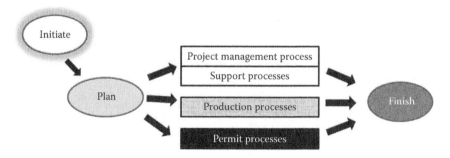

FIGURE 3.8
The main processes in a project.

- The project management process (ISO 10006, ISO 21500, PMBOK®) and support procedures.
- The production processes: how the product is produced (ISO 9000).
- The permit processes: how boards and authorities authorize and approve the execution of the project.

During the different phases, a number of subprocesses are repeated (e.g., the purchasing process). Goods and services are bought from consultants, contractors, suppliers, and others. There are also management processes, such as cost control, time control, communication with authorities, and quality control. These subprocesses are classified into different areas of knowledge. For all of them, there should be established processes based on the main project management process and baselines to monitor against. This enables working more easily with the project. The various knowledge processes are described in Chapter 6.

Describing the project flow with regard to project management knowledge areas results in Figure 3.9. The white boxes in the figure are project management processes in various areas of knowledge. There can be support processes within the knowledge areas. The budget process, for example, is a support process to the cost management process. Note that drawings and specification documents are the "products" of the design team.

3.2.1 Project Flow for DBC Construction

The project flow is divided into a number of familiar phases for builders. Activities with special requirements, such as those taking place in pharmaceutical plants and nuclear plants, require validation throughout the project. The validation flow is shown in Figure 3.10.

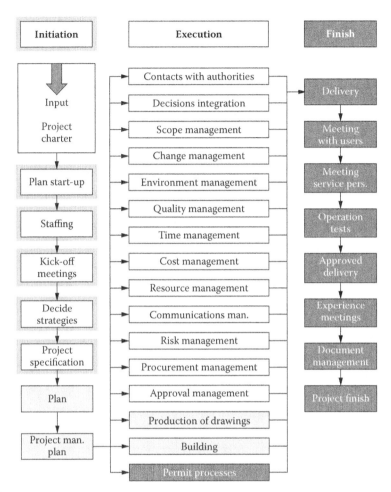

FIGURE 3.9

Project processes, including project management knowledge areas.

3.2.2 Project Flow for DC and Turnkey Construction

The project flow is divided into a number of well-known phases, as shown in Figure 3.11.

3.3 PROJECT GOALS

In describing a project, it is not enough to ask WHAT to do and HOW to do it. It is also necessary to know WHY the project should be implemented. Projects are based on needs. Projects not based on needs will end in failure.

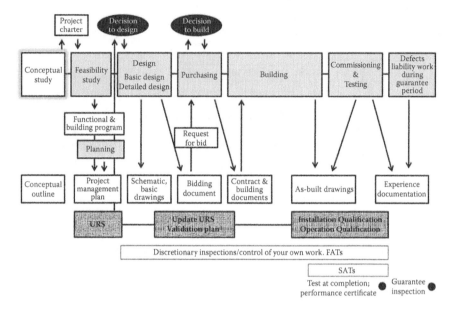

FIGURE 3.10

Project flow with documentation and validation for general construction.

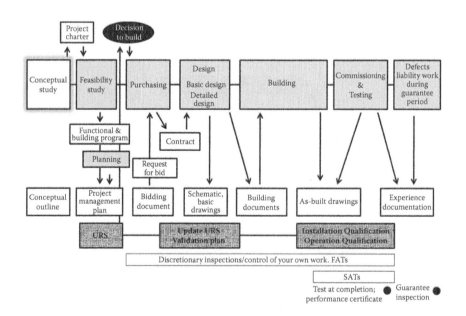

FIGURE 3.11

Project flow for DC and turnkey construction.

It is important to distinguish between needs' and objectives' requirements and wishes. Wishes are "icing on the cake/gold-plating"—something to include in the project without asking for more money and/or time. Car tires are needs, but Galaxy 500xE40 tires are most likely a wish. During the design phase, many project managers have fallen to the temptation to be a "nice guy" by approving various wishes. As an example, if, every 2 weeks at a decision meeting, a project manager (PM) accepts four wishes costing $1,500 each, after 4 months there will be a cost increase of nearly $50,000 for things in excess of requirements.

Sometimes the requirements will change during the project. The project will have to adapt, get more time and money, or close down. To close down a project is difficult, but to dispose of "good money after bad" is often not a good idea.

The project's goal is broken down and described in various parts, which are used in different ways by different project actors. To break down the project goal into functional and technical requirements is a gradual process (see Figure 3.12).

The customer can carry out a project to achieve a long-term effect, such as increased market share, improved profitability, changes in customer behavior, reduced bad environmental impact, better communications, or new product requirements. This *end-effect goal* is determined by the

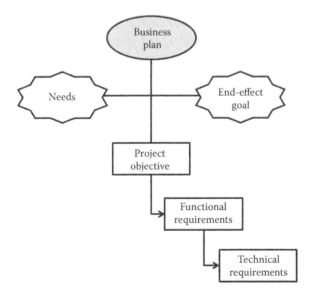

FIGURE 3.12
Structure for objectives.

project owner. The project organization must know this goal in order to understand why the project is being implemented.

Sometimes the end-effect goal is secret. A new facility, which eventually will lead to the relocation of a production plant, may be an example of a secret end-effect goal. Projects where the end-effect goal is not known or understood are often difficult to control.

The project team needs a brief and clear description of what the project should achieve in terms of scope/performance, time, and cost, as well as any special relationships that must be observed. A project can also have subtargets or milestones.

In the project's early phases (concept and feasibility phases), the project objectives are specified by functional requirements, which are studied with end-user focus. Often the architect and electrical and mechanical consultants help out in this process. The project's limitations must also be made clear and this is done in the statement of work (SOW) or a project plan. During the schematic/basic design phase, the functional requirements are transferred to a schematic drawing and technical descriptions or specifications.

The better the target is described, the greater is the chance to meet the time, budget, and quality requirements. For example, in professional bidding documents for a DBC contract, the objectives, functional and technical requirements, and solutions are normally well described. Unfortunately, there are often deficiencies that result in changes and additions. These shortcomings should be regulated according to the contract rules.

However, too often the project is not specified enough. The PM's first task is therefore to clarify the objectives, end-effect goals, functional and technical requirements, and SOW (see Figure 3.13 and Section 6.3.1).

The designers and the contractors receive additional factors in inquiry documents, bids, contracts, and their own company objectives and business plans (see Figure 3.14). In a successful project, the designers and the contractor must understand not only these documents, but also the project's functional and technical requirements.

The project needs are compiled in the conceptual and feasibility studies. The needs analysis should include a description of what the product and activities should achieve and how this will affect existing business. The needs analysis should also include a problem description, changes to the business plan, and environmental factors. During the conceptual/feasibility phase, the client or users and the PM formulate the needs (functional and building programs) that will form the basis for the next step, the planning phase. The knowledge of the project will increase during the project.

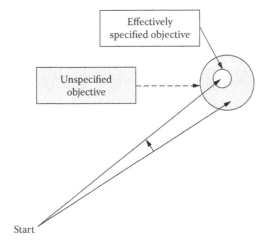

FIGURE 3.13
Specified objectives.

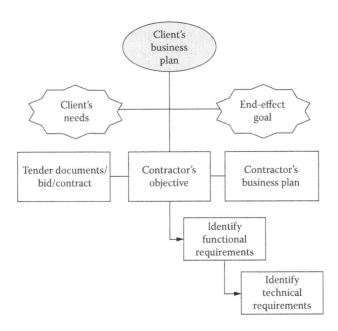

FIGURE 3.14
Objectives for the contractor.

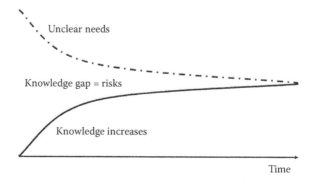

FIGURE 3.15
The knowledge gap in the beginning means not only uncertainties and threats, but also opportunities.

The difference between clear and unclear specifications—the knowledge gap—means risks (see Figure 3.15). The more badly defined the objectives are, the wider the knowledge gap is and thus the greater the risks are.

In many projects, there is a discussion about what is included and what is not included in the project. The property unit, fittings, equipment, and furniture are discussed. The property unit is land and the buildings and structures in and on the land, as well as related constructions, installations, and accessories intended for permanent use. These are generally included in a construction project.

The *built-in* fittings, equipment, and furniture tied to the particular user's activity are sometimes included in a construction project. Equipment and furniture normally mean objects, machines, and vehicles that are used in a client's business. If it is not clear in an SOW with descriptions, characteristics, etc., this must be cleared up early in the project. This can be done with provocative questions:

- Is AV equipment (projector, screen, speakers, whiteboard, etc.) included?
- Are furniture, reception desk, curtains, etc. included?
- Are refrigerator, safe, or electron microscopes included?
- Are cranes or derricks included?
- Are internal TVs, large and small computers, and cables for these included?
- Who is responsible for machinery installation, wiring, and testing?

3.3.1 Project Charter and Project Requirement Specification

During the conceptual/feasibility phase of the project, goal and objective descriptions, documentation, and approvals are successively developed (see Figure 3.16). It is in the initiation and the planning phases that **what, how,** and **when** to do things are decided.

During the initiation phase, the uncertainties in the contract are clarified. For the PM, task leader (TL), and contracts manager (CM), this study will result in good knowledge of the project and its requirements. The basic data can be an oral description, a bid, or a contract. It is in this phase that uncertainties in the order or contract are clarified in order to reduce risks and threats in the future.

3.3.1.1 The Client or Project Owner

During the initial phase, the client, together with the project teams, establishes the concept documents. The project charter (see Sections 4.1 and 4.1.1) clarifies the powers and responsibilities of the project manager. Of course, it also includes a project plan, which is a description of needs, end-effect goals, effectively specified project objective, SOW, and environmental factors. Information on the steering committee and key stakeholders is also important and should be included in the charter document (see Section 5.2).

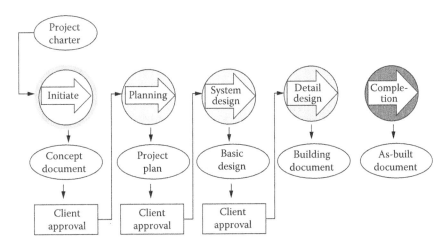

FIGURE 3.16
Design flow with various project management documents that are gradually approved.

Many times the project manager must write the project charter and have it approved by the client. During the initiation and the project planning start-up, the PM summarizes all the knowledge available in the "project specification." This is sometimes called the "preliminary project management plan." It is therefore a good idea to use the same headings as the later developed project management plan. Appropriate headings in the project specification can be

1. Approval of this project specification document
2. General
    2.1. Project name, project owner, and property ownership
    2.2. Background
    2.3. General description of needs, goals, end-effect goals, etc.
    2.4. Important stakeholders
    2.5. Strategies for different areas of knowledge
    2.6. Environmental factors and policy
3. Authorities, requirements, and permits
4. Decision integration
    4.1. Meetings with client, users, and steering group
    4.2. Preliminary model for decision making, coordination, and steering committee meetings
    4.3. Preliminary model for interface coordination
5. Scope
    5.1. Description of the project's limitations: what is and is not included
    5.2. Description of how the consecutive functional description and technical requirements will be developed, completed, and approved
    5.3. Technical specification requirements
6. Change management
7. Environment management
8. Quality management
    8.1. Audits
    8.2. Nonconformity reports
    8.3. Special tests, etc.
9. Time management, important milestones
10. Cost management
    10.1. Budget
    10.2. Cash-flow strategy

10.3. Delegation of attest authority
11 Resources management
    11.1. Human resources, special skills, organizational structure, responsibility matrix
    11.2. Computer management, software versions, CAD (computer-assisted design) guidelines, servers
    11.3. Offices, storage, workshops, etc.
    11.4. Other resources
12. Information management
    12.1. Secrecy
13. Uncertainties and risks
14. Procurement and delivery inspections
    14.1. Type of contract agreement
    14.2. Routines for delivery inspection, audits
15. Delivery and approval.

An example of a project management plan is given in Appendix A.

The project specification often looks the same as the project management plan. However, under certain headings, it may include phrases such as "to be investigated or completed during the feasibility study."

3.3.1.2 Design Consultants

The consultant manager and the client establish together a project charter for the TL. This charter includes the consultant's SOW, needs, project goals, etc. The consultant manager and TL then write a task leader charter and a consultant management plan that describe the responsibility and authority that is valid within the consultant company itself. Remember that the consultant's goals (profits, consultant business plan) are not the same as the project goals.

The consultant not only produces documents but also, in many cases, helps the client with the project management processes (e.g., scheduling and calculations and how to deal with them in the project). The consultant project SOW or specification can have the same headings as the client project specifications. Unclear points in the contract should be clarified, as well as how they should be regulated.

The following subheadings are also common in a consultant's SOW:

- Architectural concept

- Control of the results
  - Alternative design models, manual or data calculations
  - Comparisons with previously proven technology solutions
- Control of relevance of information given by client or other consultants

3.3.1.3 Contractors

Together with his or her department and the client, the CM prepares an SOW with the CM's responsibility, authority, needs, and objectives. For functional/performance contracts and design and construct contracts, the contractor's SOW complies broadly with client project specifications but will have the contractor's business plan in focus. The SOW for design–bid–construct contracts contains a lot of information in the contract documents that must, of course, be incorporated into the contractor's project management plan. To reduce the amount of text, "see contract text" is often written. Unclear points in the contract should be clarified, as well as how they should be regulated.

3.3.1.4 Transition Initiation and Planning

The project specification is an excellent communication and information tool within and outside the project team. Sometimes, however, some points must be kept confidential.

3.4 PROJECT STRUCTURE

Before the client, designer, and contractor start their work, they must obtain a work structure. If planning has been done correctly, there is a chance that the scope/performance will be met on time and within budget. But the road is not easy and there are many pitfalls.

With a good SOW and good planning, a route that leads to the goal can be developed. A common mistake is to be careless or to skip the early work that must be done. It is necessary to start right and to do this by creating a project structure. The tool to create a project structure is called WBS—work breakdown structure.

3.4.1 Create a Project Structure with WBS and Work Packages

Before designing and building begin, it is necessary to determine **what** to do. There is one plan for the entire project and one for the first part of it. In the initialization phase, the client, the designer, and the contractor therefore structure their own project. Outside the building and construction industry, project teams use WBS technology to get an idea of **what** to do and **how** to implement a project. The result is a series of work packages (WPs). In the initiation period, there is little knowledge about how to produce the building. Should it be cast in situ or prefabricated? The drawings are not made and the best method has not been evaluated. This work, which must be done, cannot be specified. It is called planning packages (PPs) and will be broken down into work packages later, when the knowledge is acquired. The WPs form the basis for scheduling, cost control, risk management, resources, etc. (see Figure 3.17).

A consultant, a DC contractor, or a DBC contractor also starts to break down the project into smaller parts, subprojects, or phases, and then further to the work packages.

In the beginning, the information about the project is small and some assumptions must be made and documented. These assumptions may later be adjusted when the WBS is done on a more detailed level (see Figure 3.18).

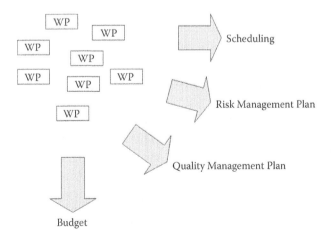

FIGURE 3.17
The work packages used in different tools and methods.

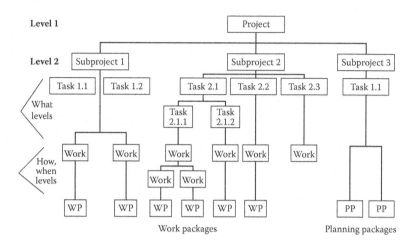

FIGURE 3.18

The WBS structure. The **what** and **how** levels can have unlimited sublevels. The subprojects can be geographical areas, buildings, or different phases of the project. The planning packages (PPs) are elements we know we must do, but we do not have the knowledge to continue the analyses. We cannot determine the production until we have designed the structure.

The WBS analyses are started by dividing the project into **subprojects** and then asking the question: **What** should we do? The responses to "what" can be broken down into smaller, more practical parts—other **what-levels.** When the *lowest what-level* is reached, the following question is asked: **How** do we do it and **when** do we do it? This is the first planning step. This **how-level** can also be broken down into lower, more specific levels. In the work packages, the information gathered is summarized and the WPs are used for cost control, scheduling, quality management, etc. The level above the WPs is the **control account**—the level for management reporting and where costs are accrued and monitored.

When the decision on **how** to carry out a task is made, **when** it should be done and **who** is responsible for doing it are also determined. **What** human and material resources are needed? **How** long a task can take and **what** the WP budget is must also be assessed. **What** must be done before the next work package can be started and **what** may start when the work package is finished? **What** uncertainties, threats, and opportunities exist in the work package?

Governing documents and supporting documents are the basis for information and planning (see Figure 3.19):

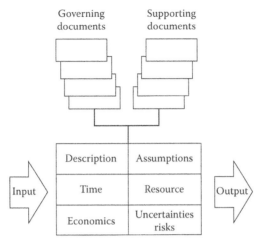

Appoint the person who is responsible for WP

FIGURE 3.19
Work package content and the documents that influence the package.

- Governing documents
  - Contract requirements
  - Public authority requirements, codes, etc.
  - Drawings, technical and other descriptions in the contract
- Supporting documents
  - Process descriptions
  - Checklists
  - Templates
  - Recommendations from trade organizations

The initiation phase concentrates on what the needs, objectives, SOW, etc., are. Whether there is a contract requirement for special approvals, production, etc., should be noted, as well as any public authority requirements and activities related to the task.

The input—that is, what must be done—needs to be described before the work package can be begun. It is also necessary to specify the assumptions made in the planning process. Whether there are any project strategies to consider needs to be stated.

What critical dates or milestones must be met? What goods (such as elevators, prefabricated frames, windows) have long delivery times and must be purchased early and perhaps separately? The last day that a work package has to be completed (internal or external milestone) should be noted. Any limitations that govern access to different areas should also be noted.

What personnel resources are needed to conduct the work? Any machines, materials, facilities, workshops, etc. that are needed should be noted.

What do the economics for the work package look like? What is the estimated cost for the work package, and what is the budget? What types of remuneration apply?

What uncertainties, threats, and opportunities are associated with the work package?

Activities that may start when the work package is finished, the output, and whether there is a lag or a lead should also be noted (see Section 6.7.2).

The name of the person responsible for the WP should be documented. Also, any requirements for information, reviews, approvals, etc., should be noted.

WBS is used throughout the project feasibility study. The tool is then used for various project phases and knowledge areas, such as design, the consultant's own work, bid work, production and parts of the production, testing, and handover.

Using a template for the work package information compilation (see the template in Appendix B) offers an early reminder to look for the governing parameters, which otherwise could be forgotten or show up late. This may result in shortcomings, which too often result in delays and/or additional costs.

3.5 PROJECT MANAGEMENT

To move from the needs, objectives, and SOW to delivery and customer satisfaction, the following are used:

- Project management processes with support processes
- Production processes
- Public authority processes

See Figures 3.5–3.9.

3.5.1 Project Management Processes

It is necessary to distinguish the differences between project management processes (ISO 10006/ISO 21500/PMBOK®) that show how to control projects and the processes for the manufacturing products (ISO 9000). See Figure 3.20.

Projects are managed with knowledge, skills, tools, methods, and processes. The main steps in the project management process are initiation, planning/replanning, implementation, monitoring/acting, and, finally, the closing process (see Figure 3.4).

During the initiation phase, what needs to be done and the objectives are studied. Then the execution is planned, and how to do it, when to do it, and who should do it are determined. During the execution, performance is monitored and compared to what is planned or the baselines. If the outcome is not satisfactory and the project needs to be set right, action is required. No project goes as planned. Therefore, it is necessary to replan, execute the new plans, monitor, and control again and again until the project is ready to be delivered.

Good project management means planning, monitoring/acting, and control of all the following areas of knowledge. For all these areas, there should be an answer to the what, how, when, and who questions for a specific project:

- Public authority contacts
- Coordination and decision making
- Scope definition and interface handling

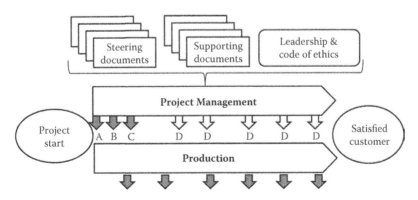

FIGURE 3.20
The project management process. A = project charter; B = project specification; C = project management plan with baselines; D = monitoring and acting.

- How to handle changes
- Environmental and occupational health
- Quality assurance
- Time
- Cost
- Resources
- Communication and information
- Uncertainties and risks
- Procurement, including administration of contract agreements
- Delivery acceptance

Some knowledge areas have support processes. One example of a support process is the budget process, which is part of the cost management process.

General Dwight D. Eisenhower said during the World War II invasion of Normandy, "Plans are nothing; planning is everything." The invasion did not go as planned; wars and projects often do not. Thanks to all the planning, the generals had great knowledge and were able quickly to establish new plans that worked, at least for a while. Planning is extremely important because it provides knowledge and the planning work is therefore not wasted.

Working with the project management processes for the different knowledge areas applies not only to the PM but also to the CM, subcontractors, managers, and the consultant team leaders. Everyone must plan and execute work in terms of time, cost, quality, resources, etc.

In many countries, the environmental and work environment hazards must be studied and a risk management plan and an emergency plan developed. How is the project team's work condition (stress) during the planning phase? How are environmental and safety issues (e.g., assembly work in tight spaces) handled during production time? How will the environment, health, and safety be affected when the product is used?

3.5.2 Strategies, Tactics, and Ethics of the Project

3.5.2.1 Strategy

Strategic issues are not pressing issues and therefore they often take second place to operational issues. However, strategy decisions are very important because they create a genuine platform for the project. Without it, speed and efficiency losses can be experienced in the form of discussions on fundamental issues for the project.

Project strategy is a long-term, comprehensive approach to achieving the project objectives. A good project has a good structure and dynamics. The strategy is needed to build the structure and manage the dynamics and it is formulated by needs, end-effect goals, and the company's business strategies (see Section 3.3). What are the objectives? The project is also governed by the stakeholders' perceptions of a number of issues. What strategies should apply to these issues?

It is important that strategies be accepted. The aim of the strategy must always be explained (see Figure 3.21). Examples of explicit strategies that are clarified early include

- Vision
- The project's technical and architectural concepts—for example, ecologically sustainable building, feng shui solutions, ultramodernism, minimum investment, reflected power (branding, city branding), or technological vanguard
- Described flexibility, generality, or convertibility.
- Whether the project should be finished as soon as possible or at as low a cost as possible

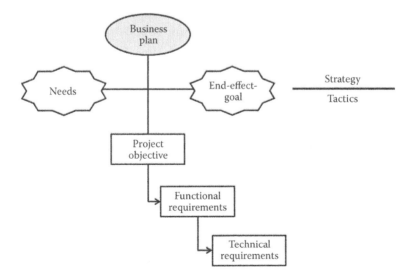

FIGURE 3.21
Strategy should be included in the project objectives and developed together with the customer and the company. The tactics are determined by the PM.

- Have a clear directive on transparency or dialogue with the public to gain acceptance
- Attitude toward client, contractor, or supplier:
  - Maximum accommodation or threatening to terminate the agreement
  - Additional invoices for all changes (the consultants, contractors, or suppliers increase their margins)
  - Interpretation of contracts in one's own favor or neutral interpretation
- Financial risk
  - The company's tolerance levels
  - Type of contract agreement
  - The preferred form of remuneration
- Alliances with other companies and suppliers of strategic goods
- Buying in parts or as whole commitment
- Specific environmental policy
- Specific energy efficiency goal for the project
- Specific health and safety policy for project execution and the use of the product
- Priority to public transport over private transport
- Whether there are ethical restrictions on purchases from different suppliers depending on, for example, health, tax, trade, or political issues
- Policy for the purchase of services and products from the company itself or its sister companies
- Personnel issues: own staff must be used before hiring outside the company
- Ethical demands in negotiations (code of ethics)
- This project's priority compared with other projects
- Determining where the focus should be on the project: time, cost, scope/performance, or a combination of two of the parameters (helpful when the steering committee and project manager take project-critical decisions)

Many of these strategies are formulated by the company and simply need to be picked up.

When, as the third project director, I took over the contract for railways across the Öresund (bridge and tunnel between Sweden and Denmark), the project was 7 months behind schedule. The client's (the Öresund

Consortium) confidence in us as a contractor had reached a low-water mark. I started by formulating the following strategy:

- Rail traffic would start on July 1, 2000.
- This meant that track, catenary, power, signal, and ATC must be completed in time for the users to test drive. In turn, this meant an earlier takeover than the contract asked for.
- Technical standards had to meet contract specifications.
- The contract should make a profit.
- We had to create confidence among the client management personnel to be allowed to give additional bids in connection with our contract.

3.5.2.2 Tactics

Together with the project management team, the PM decides which tactics to use to fulfill the strategies. What is the client's situation? What can be done to solve the client or project owner's situation? Which is the best way to reach the project's strategy successfully?

Hard parameters. When purchasing goods, it is necessary to choose strategy and tactics depending on the supply risk and impact on profit. In 1983, in the *Harvard Business Review,* Peter Kraljic described a matrix where the horizontal axis was the market's complexity (supply risk) and the vertical axis the profit impact (see Figure 3.22). This matrix identifies four different items:

- Strategic items: high impact on profit; scarcity or few suppliers; difficult, unreliable delivery

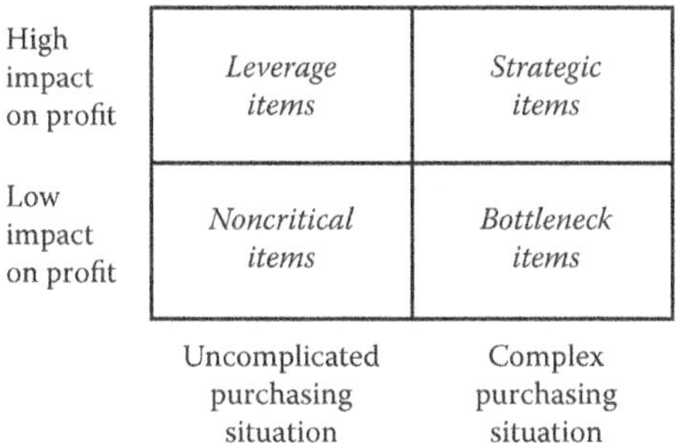

	Uncomplicated purchasing situation	Complex purchasing situation
High impact on profit	*Leverage items*	*Strategic items*
Low impact on profit	*Noncritical items*	*Bottleneck items*

FIGURE 3.22
Strategies in Kraljic's matrix.

- Leverage items: high impact on profit; many suppliers; noncomplex deliveries
- Bottleneck items: low impact on profit; scarcity; one or few suppliers; difficult, unreliable delivery
- Noncritical items: little impact on profit; many suppliers; single deliveries

The tactics for dealing with these items include

- Strategic items: balance, diversification, or exploitation
- Leverage items: exploitation of purchasing power
- Bottleneck items: volume assurance
- Noncritical items: efficient processing

In order to manage risk taking, avoid the activity, transfer the threat to someone else, reduce (mitigate) the probability and/or consequence, or accept the risk (see Section 6.11). Financial risks are managed, for example, by different types of remuneration, a clause on exchange rates, and indexation clauses.

Cooperation between the client, different designers, contractors, and suppliers leads to better results than if everyone is working only for himself or herself. No one is as smart as everyone is together.

Soft parameters. Often, focus is on the technical solutions to meet project objectives and the soft parameters stay in the background. Thinking is in terms of design, material selection, production, cost, and time. This is the tip of the iceberg. All these things rest on what lies beneath the surface—namely, *a smart description of scope and objectives, risk management,* and *relationship management.* The tasks under the surface can be just as challenging as those above. For projects to succeed, good relationships between the different actors are important. In almost all cases, technical problems are managed, but problems that are *not* clearly visible are stumbled upon.

When reviewing projects with problems, a relationship problem is almost always the root of evil when the project fails to reach its goals. A good tactic in relationship management is to create trust among the project members. Trust is the strong foundation on which the project work is based and, if it breaks down, the project decomposes.

Trust is created through respect, consideration, confidence, and cooperation (see Figure 3.23). It can be fun to go to work, even when it is tough, if a positive work atmosphere has been created. How can such a good

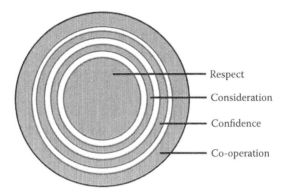

FIGURE 3.23
Trust circles.

working environment be created when people are so different? Trust must not be abused. Commercial life is a balance between relationship and trust and can influence the project cost, as shown in Figure 3.24. That old adage, "too little and too much spoils everything," is true in this case. Any party may be led to exploit a situation.

Project members come from different cultures. They come from different countries or companies. Also, different departments and regions within the same company have different cultures.

At the Öresund Bridge project, we formulated the following tactics:

- Form a functioning organization with explicit responsibilities and authorities.

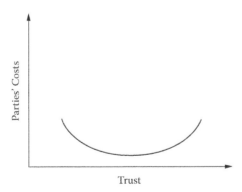

FIGURE 3.24
The relationship between trust and the parties' costs. The old adage that "too much and too little spoil everything" applies here.

- Honesty and objectivity should guide everything we do.
- Provide true information about what we do.
- Prioritize unanswered questions that irritate the client and other contractors.
- Prioritize completion date.
- $R = Q \times A$: The result of our work is the product of the quality of our work and customer acceptance of our commitment.

By having joint workshops where differences are discussed, an environment in which everyone feels comfortable working can be created. Good, straightforward, and honest relationships need to be created between employees and clients, consultants, contractors, and suppliers.

When Götaleden (a "big dig" tunnel project) was built in Gothenburg, Sweden, the following tactics were defined:

- Cultural understanding
  - Presentation
    - Introduce ourselves
    - Who we are and the roles we have in the project (two different things)
  - Culture definition
    - Identifying our cultures and business cultures
    - How we are different
  - Creating a collaboration contract
    - Formulate common goals and values
    - How we should relate to each other
  - Follow-up
    - Breaking down the goals into smaller parts
    - How the goals function in daily work
    - Rate cooperation and thereby identify weaknesses and deficiencies
    - Discuss how we can improve
- Differences and conflicts
  - Conflict management
    - Conflict resolution model
    - How we deal with disagreements with the best result for the project in focus
  - Develop tools
    - Define methods and tools to solve differences

- Joint development of improvements to these tools
- Anchor
  - Implement the methods and tools
  - How we can reach the whole team with the message
- Experience
  - Evaluate
- What we have learned and how we can teach others

Questioning can lead to something better. Can a better solution, with "win–win" for both parties, be found? However, questioning and minor differences cannot be allowed to grow into conflicts. They must be addressed immediately. Listen, show empathy (when relevant), define, and agree on what the fact is and discuss the rest. Distinguish between project problems and personal problems.

3.5.2.3 Ethics

A project manager continually has to take a position on ethical questions. There may be products produced in poor working conditions or by child laborers. These are simple issues to address. It is more difficult when it is necessary to take account of environmental issues, antitrust, negotiations, skills, or cheating. Then the PM needs someone to talk to.

Socrates argued that ethics is an interaction between questions and answers. Aristotle saw man as the "moral animal." A PM should focus on personality. What kind of person would he or she like to be? Morality is an idea of what is good and right and how to behave. Not acting as one should is called a double standard. Ethics is reflecting on and working with moral issues. Professional and business ethics are different forms of ethics in project work. What are the values in the project? For example,

- What do you do if a colleague is not doing his or her job and this means degradation in your work performance? Should you protect your colleague or speak up?
  - Try to correct the behavior and explain that there is nothing wrong with him or her as a person.
  - Ask, "Why does this happen? Do you understand that this means extra work for your colleagues? Can you do anything so that it does not recur?"

- What do you do if the boss asks you to cheat in your work in order to gain time and profit for the company?
  - Explain clearly to the boss that he or she has taken over full responsibility for quality performance.
  - Will it affect your professional pride? If so, consider whether you want to work in such an environment.
- What do you do when your tasks or the employer's actions are against the law? For example, if you suspect illegal workers, what do you do?
  - Ask your colleagues if they have the same suspicions.
- What do you do if your colleagues or managers do not follow the ethics rules that are set?
  - Ask your boss if there is a reasonable explanation for the noncompliance.
- If there is no valid reason or desire to correct the problem, you must consider whether you want to work in such an environment.

Does your trade association give you support and offer to help in dealing with ethical issues? Do you want to be involved in anything that your children will be ashamed of? What happens in your family if an ethical issue gets public attention?

If employees participate in ethical discussions in various practical issues, behavior will probably be trustworthy even in matters not discussed.

3.5.3 Control

Control is taking action when something does not follow the plans. If the different knowledge areas are evaluated and compared with what has been planned (baselines) on a regular basis, then there is control over the project. If there are deviations from the plan, it is necessary to decide on an action. The reviews are based on all the information available during the execution. Taking action helps to achieve the project goals.

Project managers control the project through decisions at meetings, as well as oral or written directives. A professional PM dares to take decisions and delegate responsibilities. *As Winston Churchill said, "The price of greatness is responsibility."*

A project manager cannot survive if he or she does not delegate. It is better for a PM to have five people working for him or her than to be doing the work of five people. When a project manager delegates, he or she still has the responsibility, but must give the delegated person the freedom of

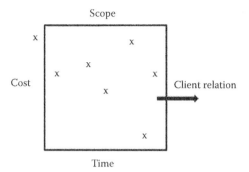

FIGURE 3.25
Devolution square. If we have delegated an operation, we should only monitor and act only if we are outside the delegation framework or it is about to fall outside the square.

action. Intervention takes place only if the person's actions have fallen outside the time, cost, scope/relationship square or are leaving the project's framework (see Figure 3.25).

Project meetings, construction meetings, schedule meetings, earned value (see Section 6.8.5), final cost estimates, control plans, risk management plans, quality, and safety meetings are the most common tools for controlling the process. All areas of knowledge must be monitored for effective control.

The PM must also periodically ask the following questions:

- Am I doing the right things?
- Are we doing the right things?
- Are we doing things the right way?
- Are we on track?
- What is essential?
- Are the objectives still relevant? Will we achieve them?
- What can we do better?

3.5.4 Tropical Projects

Some projects grow gradually as work takes place (as in a tropical forest): "We may as well include this, and this, and that…." Such projects usually grow out of control both in time and financially. The owner and the project owner should strongly consider whether the project should be allowed to grow. Maybe a new project should be undertaken instead?

Consultants and contractors have nothing against so-called tropical projects. Fixed prices very easily turn into cost reimbursement, which means a transfer of economic risk from contractor to client. The ability to control tropical projects is difficult. Time, cost, and risk processes must be repeated again and again. Employees will get tired: "Another change and I quit!" Everyone wants to see the results of their work and does not want to start over again and again.

4

Generic Main Processes

4.1 INITIATING PROCESSES

4.1.1 General

In this phase, we identify **WHAT** should be done. With the project charter as the base for the project, a smaller group develops what will be built, determines the milestones, assesses costs, identifies risks, and clarifies the basic functional requirements (see Figure 4.1).

At the very start, the project manager summarizes information and authority by writing a project charter and confirms the following important issues with the project owner or client:

- Project owner
- Project name
- Property ownership
- Project description
  - Background
  - Project's objectives, needs, and end-effect goals
- Organization model and important stakeholders
- Rough or detailed dates
- Cost frames
- Responsibility and authority

The purpose with the initiating process is to

- Clarify the project's objectives, needs, and end-effect goals
- Clarify the project/scope and performance
- Clarify work responsibilities, authority, and reporting paths of the project manager (PM), task leader (TL), and contracts manager (CM)

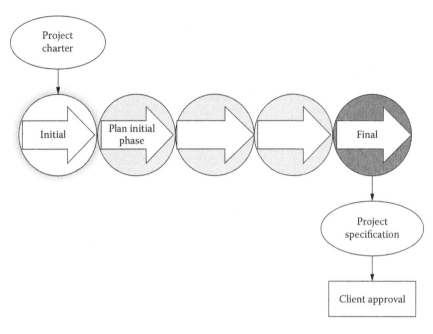

FIGURE 4.1
The flow for the initialization of various project documents should be approved.

- Clarify the client organization, end-users, steering committee, and key stakeholders
- Estimate milestones, risks, and costs
- Identify key functional and technical requirements
- Identify strategies in the relationship with the client, designers, contractors, and suppliers
- Create the conditions for the feasibility/planning study

A designer, contractor, or supplier should also clarify the order's or contract's restrictions.

The initial study is summarized in a project specification. This is the precondition for moving forward with the feasibility study. The PM's authority and responsibility, as well as needs, end-effect goals, and a "SMART" (specific, measurable, accepted, realistic, time specific) objective statement should be included (see Section 3.1.2 in the previous chapter).

A DBC (design, build, construct) contract or a request for a design bid is, for a designer, the foundation of the project. In order to start the assignments right and get an easier job situation later, the designers and contractors must study these documents scrupulously.

Regardless of whether a person is a PM, TL, CM, planner, or foreman for a contractor, it is important to start right. Determine as soon as possible (see Section 3.1.2) the objective of the sub-project (e.g., create a basic design, erect a superstructure, purchase components, perform tests). Examples of initiations in the construction industry include the following:

Actor	Activity	Deliverable
Project owner	Determine PM responsibility and authority	Project charter
Consultant	Study request for bid; estimate and negotiate	Bid for design work and contract
CM	Study request for bid; estimate and negotiate	Bid for contract work and contract
TL	Study contract regarding time, budget, SOW, risks, recourses	Baselines to control design work
Foreman for contractor	Study contract regarding time, budget, SOW, risks, recourses	Baselines to control production

A PM's responsibility and authority should be identified and can look like this:

- Responsibility
  - Be the project contact person in relation to the project owner, client, or chairman of the steering committee.
  - Identify the needs, objectives, end-effect goal, and constraints and describe them—"SMART."
  - Plan, monitor, and control the project against the baselines.
  - Plan the project, carry out a feasibility study, and prepare a project management plan that can be approved by the client and the steering committee (if any).
  - Identify project stakeholders and keep them informed.
  - Maintain required project documentation.
  - Report project status to stakeholders in accordance with the communication management plan.
  - Immediately report to the client if events, risks, or follow-ups indicate that targets will not be met.
- Authority
  - Represent the project in all aspects in relation to the client or project owner and other stakeholders, the latter in accordance with the project charter and the project management plan.

- Be responsible for the project budget and have the power to approve or attest to all project expenditures. This item can be limited—for example, "For orders and payments over $XXX,XXX, the project owner/department manager should be informed and (if necessary) approve the order/payment in advance of such implementation."
- Reports to
  - Client or project owner
  - Chairman of the steering committee

Many times, responsibilities and powers are specified in an overall organizational plan or a company project manual. If so, the PM should refer only to this document and supplement it with any discrepancies.

4.1.1.1 Cooperation with the Steering Committee

In my opinion, the chairman of the steering committee should coordinate and direct the work of the steering committee. Too many project managers have used up their confidence by trying to unite the steering committee members when there is a power struggle between different interests. It is appropriate that the PM sit in on the steering committee meetings, at least part of the meeting, to clarify and inform on the current status of various issues. Steering committee members must not inform project members and stakeholders about decisions taken in the committee because that is the PM's task. A second line of command can be a disaster. If a stakeholder contacts someone on the steering committee, the committee member should refer the stakeholder to the PM and possibly promise that the steering committee will ask how the PM dealt with the matter. Any decision must be communicated through the PM.

The project manager should have all decision-making power delegated to him or her. If the PM mismanages the power, it is the project owner's or department manager's responsibility to act.

4.1.2 Preliminary Activities for the Whole Project

The PM must work together with the project owner to formulate the project charter (see Sections 3.3.1 and 4.1.1). The project specification could very well have the same headings as the upcoming project management plan (see Section 3.3.1 and Figure 4.2). Some sections may include phrases

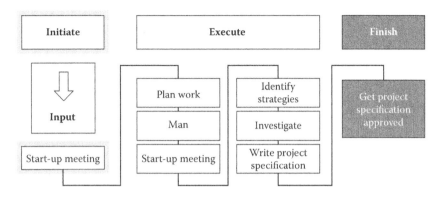

FIGURE 4.2
Project specification process. The input is the project charter and project plan

like "to be investigated/or completed during the feasibility study." If the specification is distributed to project members and appropriate stakeholders, consistent information is given to everyone. Experience shows that distributing required information to project members and stakeholders in this way is much better than verbal reports from the meetings.

4.1.3 Initial Activities for Consultants

If the consultant has a contract, it is to be hoped that it will describe the project scope, remuneration, intermediate milestones, and time for completion. Many consulting tasks are not the result of an offer and a contract. Often, the consultants start work during the initiating phase and the statement of work (SOW) is clarified gradually. At this stage, the TL will have to help the client to formulate the SOW.

Regardless of how the commission was ordered, the TL at the consulting firm needs to clarify the power granted to him or her within the company. Therefore, both *a contract with the client* and a *project charter for the TL* must be established (see Section 3.3.1).

The TL's task must be staffed, strategies defined, and the work planned (cost, time, and resource and risk management). The TL has to summarize all information in his or her task specification (compare to project specification; see Section 3.3.1). The TL must also develop the WHAT, WHEN, and HOW for his or her task's knowledge areas and summarize this in baselines and the task management plan—the consultant's project management plan. The consultant will gather information from and can refer to the client's project management plan. The designer's production (i.e.,

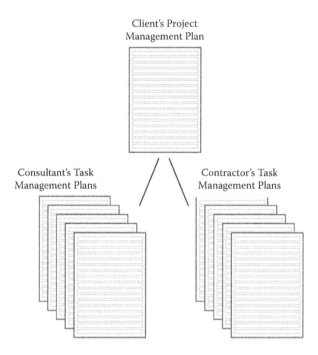

FIGURE 4.3

The consultants' and contractors' project management plans with information on all knowledge areas.

production of documents) is controlled according to the company's quality management system and coordinated with the client's quality management program. See Figure 4.3 and compare Figure 3.20.

The designer is often asked to help the client to describe the *what*, *when*, and *how* for some of the project's areas of knowledge (part of the client's project plan). This may be, for example, the product's characteristics and constraints; project scheduling; regulatory requirements; environment, health, and safety; quality control; and risk management. These are common types of side orders.

4.1.4 Initial Activities for Contractors

Sometimes the contractor gets his bid–construct or turnkey order through successful sales without competition. The initial stage is then comparable to the consultant's initial phase described earlier. However, most contracts begin with a request to be answered. "Should we bid?" Someone must be commissioned to prepare a bid.

In properly controlled projects, a project charter for the bidding work exists. How much is the cost of the bidding process? Are there any special strategies for bidding? Who can order external consulting help and quantity calculations, if they are needed? When must the offer be delivered? What threats and opportunities are in the work? What should builders do and what should subcontractors do? Who will approve the bid before it is submitted?

Before a contract is negotiated a CM must be appointed. The CM must sit in on the negotiations to get the "unwritten understandings" in the contract. The contractor's project is divided into subprojects, manned and controlled as recommended in this book. The CM should develop the WHAT, WHEN, and HOW for his or her knowledge areas and summarize them in baselines and the contractor's management plan (compare to project management plan; see Section 3.3.1). The contractor should obtain information from and can refer to the client's project management plan. All areas of knowledge (see Chapter 6) must be planned, monitored, controlled, monitored, controlled, replanned, monitored, etc.

Depending on contract terms, certain parts, such as product specifications, are already provided. Strategies and boundaries between the contractor's own work, the client work, and subcontractors' work, which were stated or assumed in the bidding stage, must now be clear and unambiguous.

The structure, tools, and methods presented in this book are general and can be used by changing those things that need to be changed through the contractor's control of the contract.

4.2 THE PLANNING PROCESS

4.2.1 General

Abraham Lincoln is quoted as saying,

> *"If I had eight hours to chop down a tree, I'd spend six hours sharpening my ax."*

Before the PM, TL, and CM start, they must plan *how* and *when* to perform their own work. They must obtain a structure of work (see Section 3.4) and decide on processes and milestones for the production and all knowledge areas. They must prepare baselines against which to monitor the work (see Figure 4.4).

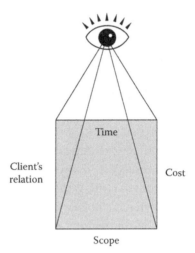

FIGURE 4.4
Project control.

A reference plan contains many parts and baselines, the most important of which are schedules, quality control plans, production-based budget plans, the resource plan, procurement plan, and risk management plan. The reference plans are compiled in the project management plan.

During the implementation phase, work is compared with the schedule and costs compared against a procurement plan and the production-based budget plan. The results of quality control are analyzed, identified uncertainties (threats and opportunities) are monitored, weak client relationships detected through "management by walking and listening" (MBWAL), etc. MBWAL means talking to one's own and other employees in the project—not just with submanagers and the client's project leader—and listening to what they have to say. There are many protective filters on the information path up to the manager.

The purpose of the planning process is to

- Specify product requirements further
- Establish reference plans or baselines for all knowledge areas
- Design reference plans or baselines in such a way that they can be monitored against during the implementation phase
- Produce the necessary baselines for controlled monitoring during the production phase

"Failure to Plan is to Plan for Failure."

—Winston Churchill

We plan by asking questions:

- **What** is necessary to know in each knowledge area?
- **What** needs to be done in each knowledge area?
- **When** and **how** can the information be obtained?
- **When** and **how** should these tasks be performed?
- **Who** will provide the data or does the job?

A good tool to develop these controlling documents is the work breakdown structure (WBS) (see Section 3.4.1). In the planning phase, it is important to distinguish between what is *important* and what is *urgent*. The planning is compiled in the project management plan, which can have the following headings (see also Section 3.3.1):

- Project management plan/task management plan/contractor's management plan approval
- General
  - Project name, project owner, client, property ownership
  - Background
  - General description of project goal
  - Specific strategies
- Regulatory contacts with authorities
- Processes for decision, coordination, and integration
- Scope management and product specifications and interfaces
- Change management
- Environment and health management
- Quality management
- Time management
- Cost management
- Resource management
- Information and communications
- Risks/uncertainty management
- Procurement
- Project approval

4.2.2 Planning Processes and Knowledge Areas

To get the structure in the planning work, the planning flow shown in Figure 4.5 can be followed. At the beginning, alternative designs are

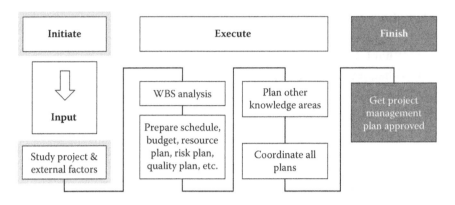

FIGURE 4.5
The planning process. The input is the project charter and project specification.

studied and checked against available time and resources; whether the execution cost is within budget is determined. It is also important to perform the risk analysis process of the production for each part to detect opportunities and threats that exist and whether they can be accepted. In connection with the execution plan, from the beginning, *how* and *when* to monitor the work progress must be noted. The planning processes for the different knowledge areas are described more fully in Chapter 6.

4.2.2.1 Contacts with Authorities

- **Purpose**
  - Ensure that the planned project is implemented in accordance with relevant laws.
  - Ensure that the right building codes are used.
  - Determine the public authority contacts required for this project and who has the authority to represent the client or company.
  - Determine and ensure adequate time for authorities' handling of review.
- **Questions to ask**
  - **What** building codes apply?
  - **What** needs permission and approval?
  - **How** and **when** can these permits be obtained?
  - **Which** agency contacts are required for the project?
  - **Who** may represent the client or company?
  - **How** long will the handling of the authorities take?

- **How** can this be discovered?

Plan and inform!

4.2.2.2 *The Process of Decision, Coordination, and Integration*

- **Purpose**
  - Ensure that decisions are taken at the appropriate level.
  - Clarify the steering committee's overall decision-making level and assure regular meeting times.
  - Clarify the powers of the PM and other subleaders and at which meetings a decision can be made.
  - Ensure that the integration is an ongoing activity.
  - Ensure that the interfaces between ongoing project, production, and maintenance divisions are handled in an acceptable way.
  - Ensure that these interface problems are solved in a structural way.
  - Ensure that meetings are efficient and deal with the important issues.
  - Ensure that only issues important to meeting members are discussed.
  - Ensure that the right meeting members come to the right meetings.
  - Ensure that the meetings are planned at such a time that those who should attend can plan other activities in such a way that they can attend.
- **Questions to ask**
  - **What** are the powers of different project participants? (Compare Section 4.1.1 in Chapter 4.)
  - **What** decision is needed to implement the project?
  - **What** is needed to coordinate the project and which client activities are affected by the project?
  - **What** is known about the client's gradual transfer of existing spaces and facilities?
  - **What** is known about the client's gradual takeover of the project?
  - **What** is known about the client's operational capacity of water, heat, and electricity and can it be used during construction?
  - **How** and **when** can this information be obtained?
  - **How** and **when** will it be established and documented how the coordination should be done to minimize mistakes and disruption of the client's ongoing activities?
  - **How** and **when** are this company's ongoing activities integrated with the project's activities?

- **How** and **when** are the client company's and users' activities integrated with this project?
- **How** and **when** will access to the premises be available?
- **How** and **when** can existing services be used for temporary water and sewage, heating, and electricity?

At the beginning of a project, it is necessary to understand how decisions are taken and how to establish this in the client organization. *When* does the board decide to implement the project and what documentation do its members need for a decision? *Who* makes decisions along the way to the finished building/construction documents and who decides which type of contract to choose? *What* powers to make decisions does an architect or a building inspector have? *How* is information about future end-users, etc. obtained?

The PM must plan the decision meetings, as well as design and building coordination meetings, and describe the responsibilities and powers in connection with them. The contractor must schedule the meetings with subcontractors, security organizations, unions, suppliers, etc.

Plan and inform!

4.2.2.3 Scope Management

- **Purpose**
  - Determine what is and is not included in the project.
  - Ensure that functional requirements are established and documented.
  - Ensure that specific technical requirements are established and documented.
- **Questions to ask**
  - **What** are the functional and technical requirements?
  - **What** is included in the project and what is not?
  - **What** can be forgotten in the interface between this project and other projects?
  - **How** and **when** can information about the requirements in the interfaces be obtained?
  - **How** and **when** will it be established and documented how the coordination should be done in the interfaces?
  - **How** and **when** will routines for information exchange in the interfaces be established?

A common problem in the project approval at completion is that unexpected things, which therefore are not in the budget or on the schedule, must be included. To avoid this, certain questions must be asked in the beginning. For example, "Is the PC projector in the conference room included in this project or in another project? Who unloads and transports the machines into the factory plant? Who assembles computer wiring and plugs for the intranet and computers? Who should be involved in determining the local characteristics? How should security be handled? Have the operational and maintenance departments taken part in the scope process?" There must be a process to follow to include all the needs but keep the wishes out because they ruin the budget. Plan how this should be done.

Plan and inform!

4.2.2.4 Change Management

- **Purpose**
  - Ensure that changes and additional work are based on real data.
  - Ensure that changes and additional work meet functional and technical requirements or that the client understands that the changes and additional work have an effect on the original requirements.
  - Ensure that risks and cost and time effects are studied before making the change.
  - Ensure procedures to document cost and time changes due to scope changes or additional work.
  - Ensure that decisions on changes are taken at the appropriate level.
  - Ensure that information about changes is spread quickly to everyone in the project organization.
  - Ensure that the change is carried through.
- **Questions to ask**
  - **What** are needs and **what** are wishes in a change request?
  - **How,** by **whom,** and **when** can a decision be taken?
  - **How** will a change influence the functional and technical standard, time, cost, risks?
  - **How** is everyone informed about the approved change?
  - **When** will the change be implemented?

Can users, architects, construction managers, or mechanical inspectors make changes without approval of the PM? Many projects have had trouble because changes were made without a proper approval. Note that this does not mean that the PM must decide everything. The PM should delegate. It is necessary to plan how the change process should be conducted. There must be a process that ensures that costs, time, risks, function, etc., are considered before decisions are made.

Plan and inform!

4.2.2.5 Environment and Work Environment Management

- **Purpose**
  - Ensure that key environmental aspects are identified.
  - Ensure that the client's environmental targets are identified.
  - Ensure that the activities to achieve environmentally better solutions are described.
  - Ensure that key safety issues are identified.
  - Ensure that the client's work environment goals are identified.
  - Ensure that environmental and work environment issues during construction and for the final product are managed during the design.
  - Ensure that health risks are minimized. This applies to the project as well as the product and its operation and maintenance.
- **Questions to ask**
  - **What** is environmentally not acceptable or not desirable?
  - **What** are the environmental standards and targets?
  - **How** can environmental requirements and environmental objectives be determined?
  - **How** can the laws and OSHA regulations that apply and the needed permits be determined?
  - **How** are the environmental routines in the project determined?
  - **Who** is responsible for environmental issues in the project?
  - **What** work environment is not acceptable or not desirable?
  - **How** are the work environment routines in the project determined?
  - **How** is responsibility delegated?
  - **How** is the fire protection organized for the project during the production (e.g., when welding, cutting, or heating is taking place)?
  - **Who** is responsible for SHE (safety, health, and environment) in the project?
  - **Who** is responsible for SHE questions in the design work?

- **Who** is responsible for SHE coordination during building/construction?
- **Who** is responsible for SHE for subcontractors' work?

Environmental and safety planning is an important project component. The risks to humans or the environment or of business damage must be minimized. SHE issues must be planned for. The authorities have set minimum requirements and many companies have additional requirements that must be met.

Plan how the SHE work will be integrated into the project work!

4.2.2.6 Quality Management

- **Purpose**
  - Ensure procedures to control the work and for reviews and approvals.
  - Identify and secure the key requirements and expectations to get the client's or project owner's approval.
  - Ensure identification of areas with validation requirements.
  - Ensure processes for areas with validation requirements.
  - Ensure procedures for receiving and inspection of goods.
  - Ensure procedures for final approval of the project.
  - Ensure procedures for experience feedback and improvement of the work.
- **Questions to ask**
  - **What** should be delivered (functional and technical requirements)?
  - **What** quality assurances should be obtained from suppliers?
  - **How** can *delivery* of the right quality be ensured?
  - **How** can *receipt* of the right quality from suppliers be ensured?
  - **How** can redoing and fixing be avoided while at the same time developing one's own work methods?
  - **How** can temporary work that meets needed requirements be ensured?
  - **How** can the project participants engage themselves in controlling the functional and technical requirements?

The railway industry, the road industry, the nuclear industry, etc., have rules about how quality control should be performed for things such as casting, welding, electrical systems, and manufacturing of pressure vessels. But the projects normally need further checks. Discretionary inspections and control of one's own work, incident reporting, quality audits, quality meetings, factory acceptance tests (FATs), and site acceptance tests (SATs) (see Sections 6.12 and 6.13) are obvious components in the quality work.

Plan and establish project-specific quality plans, control plans, and audit processes (when necessary)!

4.2.2.7 Time Management

- **Purpose**
  - Ensure that the work is done in the correct sequence.
  - Ensure that long-lead-time items are ordered and delivered in time.
  - Ensure that dates for review and approval are set and that a necessary time frame is available for this work.
  - Ensure that important milestones for the project are identified and well known.
- **Questions to ask**
  - **What** are the project's time requirements?
  - **What** long delivery and turnaround times exist?
  - **What** are the critical activities to get the project finished on time?
  - **How** can it be ensured that work is progressing in such a way that the project can be finished on time?
  - **What** activities must be adjusted to compensate for "lost" time? **How** can this be done?
  - **How** can uncertainties in the time estimates be addressed?

Before a task can be started, it is necessary to identify what must be done. It is important that the foundation for the superstructure be complete and strong enough before erection of columns and beams. What is the estimate of how long a task will last and when must it start or be finished? When are the approved drawings to start the building work needed so that the project can be delivered on time? When a schedule is used, it is possible to check whether work is ahead of or behind the planned work. Will the project be ready in time? There are several scheduling tools—for example, milestone charts, Gantt charts, and network diagrams (see Section 6.7.1). These are the most common tools for guidance on and information about project performance.

Schedule the entire project, the design work, the production, and the approval processes. Establish schedules that can be used when monitoring the work!

4.2.2.8 Cost Control

- **Purpose**
  - Ensure that the project's costs are estimated before decisions on implementation are made.
  - Ensure that actual costs for ongoing work can be monitored against the established budget.
  - Ensure cash flow and identify potential requirement of loans for the project and when this is needed.
  - Ensure procedures for invoice approval and invoice payment.
  - Ensure that financial records, securities, insurance documents, etc., are stored and handled properly.
  - Ensure procedures to follow up costs for changes and additional work.
- **Questions to ask**
  - **What** cost constraints does the project have?
  - **What** is the estimated cost?
  - **What** types of remuneration should be attempted?
  - **How** much will be paid and when?
  - **When** do salaries, supplies, subcontractors, etc., need to be paid? **How** much?
  - **How** much must be borrowed and **when?**
  - **How** can a control be developed to ensure that the work is progressing in such a way that it is within budget?
  - **How** can uncertainties in cost estimates be addressed?

The Swedish cartoonist Staffan Stolle wrote a long time ago, "Here they will build for 10 million dollars. I wonder how much that will cost."

A project that does not end within the project budget can end in bankruptcy for the client and/or the contractor. Project managers must have control over the economical development of the project. Budgets, estimated budget at completion, and tools such as earned value (see Section 6.8.5) can help to study the development and offer a chance to act if necessary. One can be forced to compromise on scope or be prepared to invest more money in the project. The estimated costs are often based on parameters like square meters, cubic meters, watts, number of drawings, etc. These figures cannot help in getting information about the economic progress. They cannot be monitored against.

Plan the work and develop the output-based budget, cash flow, cost controls, and verifications!

4.2.2.9 Resource Management

- **Purpose**
    - Ensure that the organization has the right human resources at the right time.
    - Ensure that the right equipment and machines are available at the right time.
    - Ensure that all project members work with the correct versions of the software used in the project.
- **Questions to ask**
    - **What** human resources are needed?
    - **What** key skills and experts are required?
    - **What** machines, equipment, storage, workshops, and office facilities are needed?
    - **What** are the various project members' responsibilities and **what** powers do they have?
    - **What** can be done to get participants to work toward the same goal?
    - **How** can cooperation between various participants be facilitated?
    - **How** can project members develop their skills in the project?
    - **How** can questioning be dealt with and conflicts avoided?
    - **When** are different resources needed?

What human and mechanical resources are needed? Are experts needed? Are appropriate levels of competence and capacity within the project? What software versions are used? What are the needs for the subcontractors' offices and workshops at the construction site?

Plan the project's and subproject's resources!

4.2.2.10 Communication Management

- **Purpose**
    - Ensure that the right people get the right information at the right time.
    - Ensure that progress reports are sent to the right people.
    - Ensure that the filing is done correctly.

- Ensure that digital documents and operating instructions are carried out to the client's satisfaction.
- **Questions to ask**
  - **What** needs to be communicated and who should have the information?
  - **What** needs to be archived?
  - **What** are the legal and contractual requirements for information and archiving?
  - **How** and **when** should information about different things be conveyed?
  - **How** is archiving to be achieved?

Scheduling and following up on costs against budgets are well known in our industry. However, experience shows that communication between project participants and to stakeholders is poor because the process has not been identified. My experience is that more than 80% of all problems in a project have their roots in poor communication, which creates strained relationships. When did the construction industry not solve a technical problem?

Plan for the information, communication, and archiving.

4.2.2.11 Risk and Uncertainty Management

- **Purpose**
  - Identify events that can have a positive or negative effect on the project or product.
  - Analyze and prioritize events.
  - Ensure activities so that the project finances, time, or scope/performance is not compromised more than can be accepted.
  - Ensure that there are necessary reserves available.
- **Questions to ask**
  - **What** can go wrong in the project?
  - **What** can go better than planned?
  - **What** are the ambiguities and uncertainties?
  - **How** can the risks that threaten the success of the project be reduced?
  - **How** can opportunities to get a result that is better than planned be increased?
  - **How** can the uncertainties that must be accepted be handled?

- **When** should actions to minimize threats and maximize the opportunities be undertaken?
- **When** should response activities for the project and parts of it be identified, analyzed, and planned?

All projects contain uncertainties, both threats and opportunities. How does one deal with these risks? For many years, the building industry has tried to transfer the risk to the other party in a contract agreement. Now, those who take the risks want to get paid for doing so. What is the price for adding reserves to bids, budgets, etc.? If too much is added, there will be no project and no contract. Are there other ways to handle uncertainties? It is necessary not only to deal with the uncertainties and risks of the finished product, but also to carry the risk management process during the project. What are the threats to meeting the day of completion as contracted, keeping costs within budget, and delivering as contracted?

Risk management processes must be planned for the entire project, parts of the project, or special activities!

4.2.2.12 Procurement Management

- **Purpose**
  - Ensure that the right things are bought and delivered to the right place at the right time.
  - Ensure that prospective contractors have adequate resources and can meet the requirements for quality, environment, and safety.
  - Ensure that the supplier has the financial capacity to implement the order.
  - Ensure that business is carried out in an ethical manner.
  - Ensure that scope, responsibilities, finances, time, patent protection, confidentiality, and dispute management are clear.
  - Ensure that standard terms and conditions are used when contracts are written. If not, ensure that texts of agreements are well studied, well written, and understood.
  - Ensure that manufacturing, construction, installation, and commissioning meet contract requirements.
- **Questions to ask**
  - **What** should be bought?
  - **Which** form of delivery should be chosen?

- **Which** form of collaboration should be chosen?
- **Which** form of procurement should be used?
- **What** should enquiry documents look like?
- **How** can it be determined whether a prospective supplier can deliver the right quality and on time?
- **How** will the price be determined and the seller be paid?
- **How** can the delivery process be gradually followed up and supplies be ensured to meet the contract requirements?
- **How** can the administrative requirements of the contract be met?

The purchasing process includes three parts:

- Prior to bidding
- Biddig, evaluation, negotiation, and contract writing
- Contract administration

What potential suppliers are there? How does one check that they meet requirements? How will the bidding and procurement be done? What standards should be used? How will the procurement be monitored during the implementation phase?

Plan purchases and delivery controls!

4.2.2.13 Contract Approval Management

- **Purpose**
  - Ensure delivery acceptance.
- **Questions to ask**
  - **What** can be done to get the project approved?
  - **What** regulatory inspections are needed?
  - **What** documentation and training are included in the contract?
  - **What** administration should be done in connection with approval?
  - **What** does the approval process look like?
  - **How** can errors and deficiencies identified in final inspections be corrected?
  - **How** can the administrative requirements of the contract be approved?
  - **How** much time is needed for final inspection and approval?

Many standard forms of contract (e.g., FIDIC forms) have instructions for "test on completion and client's taking over." Sometimes, it is necessary to supplement with FATs and SAT 1, SAT 2, SAT 3 to ensure that the product is finished on time and does not contain errors and defects to be corrected at a late stage, with delays as a result. See Sections 6.12 and 6.13.

A PM wants to be able to submit the project in such a way that clients and users will be satisfied. He or she must therefore schedule meetings with users, managers, and operations and maintenance departments to ensure that the agreed documentation for operation and maintenance is submitted and that training is done. Performance security or surety bonds must be written down or returned. Remedying of defects must be carried out and approved. This is easy to forget when the next project already has begun.

Plan early, agree on how to test, test upon completion, and the client's taking over will be done!

4.3 PRODUCING PROCESSES

4.3.1 General

- **Purpose**
  - Ensure that the scope and its performance are delivered on time and within budget.
  - Ensure that everything is "done right the first time" to avoid rebuilding and remedy defects.

See Figure 4.6.

4.3.2 Design and Engineering

Project designers and engineers often have two tasks. One is to produce documents that describe what will be manufactured and how to build it. But this is not enough. The documents must also make clear what is included or not included in the contract when products and contractor's or subcontractor's work are bought. The drawings and technical descriptions are also legal documents.

The other task is to help the client with management tasks—for example, specifying functional and technical requirements, scheduling, cost

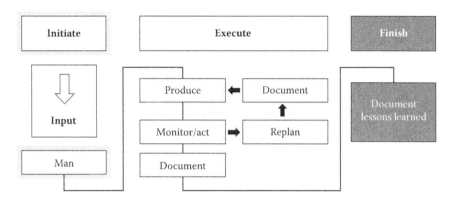

FIGURE 4.6
The execution process.

calculations, and development of quality and inspection plans for the project. Drawings and technical specifications should be made so that in the future, after completion during the contract period, they can be used for operation and maintenance.

The engineers often help the client with the following project management documents:

Administrative requirements	Knowledge area: procurement management
Outline/layout site plan for contractor's disposal of land and access route	Knowledge area: procurement management
Safety, health, and environmental plan	Knowledge area: environmental management
Project-specific quality plan and/or special project-specific inspections	Knowledge area: quality management
Risk management plan	Knowledge area: risk management

Production of documents should follow the ISO 9000 processes, but the consultant's project management process for production of these documents must be planned. This is done by preparing processes for all relevant knowledge areas described in this book.

4.3.3 Production

The main contractor buys or rents its own factories, manufactures products, and transports them to the site. On the site, the product that the customer ordered is built and installed. Material suppliers, carriers, and subcontractors help the contractors. The main contractor then becomes a customer with demands for planning of purchasing, receiving inspection, etc.

The choice of production method is and should almost always be the contractor's responsibility. He or she knows the knowledge and resources available within the company and what the different methods will cost. This is one of the contractor's most important competitive tools.

Manufacturers, contractors, and installers normally follow the ISO 9000 process, but must also control their own project management processes by planning the knowledge areas described in this book.

4.4 TESTS ON COMPLETION AND CLIENT'S TAKING OVER

- **Purpose**
  - Ensure that the agreed delivery meets contract requirements.
  - Ensure that the delivery is approved.
  - Ensure procedures during the warranty period.

See Figure 4.7.

Compared with many other industries, such as the computer industry, good guidance (see FIDIC) is available in the building/construction industry for the client's taking over. Various types of normative inspections, coordinated testing, and defect liability periods are well structured by contract standards. Yet the client and the contractor still have claims, disputes, and arbitration. Why? The client does not find that the plant or

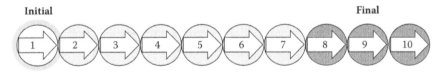

Initial   Final

1 — 2 — 3 — 4 — 5 — 6 — 7 — 8 — 9 — 10

1. Discretionary/self-inspections   5. SAT 1   8. Test on completion
2. Normative inspections   6. SAT 2   9. Remedying of defects and inspections
3. Before cover-up inspections   7. SAT 3   10. Guarantee inspections
4. FAT

FIGURE 4.7
A client's taking over the process.

product is flawless when it is taken over. Inspections and test work take time, but normative inspection FATs and SATs can be planned in a way so that the defects are gradually corrected and the remaining points at test on completion will be few.

4.5 PROJECT CLOSE OUT

- **Purpose**
  - Ensure that relevant documents are archived.
  - Ensure feedback.
  - Ensure that financial matters and the guarantee end in a proper manner.
  - Ensure that the user and the operating organization have the necessary information and training.
  - Ensure that warranty issues can be handled properly during the defect liability periods.
  - Clarify who is responsible for the financial reserves that should exist for the project to handle deficiencies discovered during the defect liability period that are not warranty issues.

If a PM fails in delivery of the project to the project owner, operating organization, and users, the entire project will be judged as unsuccessful. If there is a bad transfer and the education is poor, it is often because of lack of documentation and information.

The close out is delicate work that needs to be disciplined. Most project participants have already moved to the next project. Documentation, such as operating and maintenance instructions, as-built documentation, CE certificates, etc., is often done in an unengaged manner. The training of operating and maintenance personnel is often overlooked. Procedures for dealing with warranty issues, names, phone numbers, etc., are often lacking.

Feedback about experience that is communicated to the client, consultants, and contractors must be performed in a way that the building/construction industry can develop positively.

Plan the process of how this should be handled at the beginning of the project.

5

Project Phases

5.1 INTRODUCTION

Industrial construction projects follow certain phases in which the consultants, suppliers, and contractors are involved:

- Project initiation/conceptual study
- Project planning/feasibility study and programming
- Project implementation
  - Design
  - Building
  - Testing
  - Guarantee period

See Figure 5.1.

5.2 INITIATING PROCESSES/CONCEPTUAL STUDY

"It doesn't matter how fast you're going if it's in the wrong direction."

—Stephen Covey

The project initiation/conceptual study begins with a project charter (PC) and a project specification/preliminary project management plan. The PC describes the project manager's responsibilities and power as well as the power of the steering committee. A small group, together with the project manager (PM), will develop the project specification/preliminary project management plan, study conceptual times and costs, identify risks,

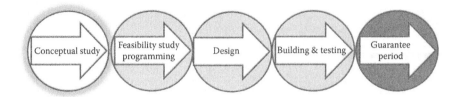

FIGURE 5.1
Project phases.

and clarify the basic functional and technical requirements (see Section 4.1 in Chapter 4).

- **Purpose**
  - Clarify the PM's responsibility and authority.
  - Clarify project objectives, the user's requirements, and estimated time, costs, and risks of the project.
  - Obtain sufficient information in order to have an approval of further work on the project.

This phase also includes the development of a preliminary project management plan for all the knowledge areas. The presentation of the conceptual study should therefore also include the following control information or an explanation as to why the information is missing and when and how it will be studied:

1. Information on who prepared, reviewed, and approved the document
2. Project summary description
   2.1. Project name, project owner, client, property ownership
   2.2. Background
   2.3. Project objectives and end-effect goals
   2.4. Product specification
       2.4.1. Overall functional requirements
       2.4.2. Special technical requirements
   2.5. Conditions for project approval
       2.5.1 For phases and overall approval
   2.6. Project strategies
3. Contacts with regulatory authorities
   3.1. Building codes and permissions and approvals needed

4. Decision, coordination, and integration
    4.1. Approval of documents
        4.1.1. Project charter
        4.1.2. Project specification
        4.1.3. Preliminary project management plan
    4.2. Who has the right to make decisions on different issues and how should these be anchored?
    4.3. What documents and phases must be approved before continuing to the next phase?
    4.4. How should the project be integrated with ongoing activities?
5. How should change management work?

For help, an FBS (functional breakdown structure) can be used (see Section 6.3 in Chapter 6). Compare this with a WBS (work breakdown structure). In the FBS, questions are asked about functional and technical requirements, as well as how and with which specifications these functions will be met (see Section 6.3.1).

5.3 PROJECT PLANNING/FEASIBILITY STUDY AND PROGRAMMING

Project planning/feasibility study and programming is the planning phase for the project. In this phase, the project is further specified with the project specification as a base. This is before preparation of systems and basic design documents begins. Through interviews, field visits, key ratios, new risk analyses, and cost and time estimates, a project management plan is developed that should be the basis for the first step of the project implementation: the preparation of the drawings and building specifications. The architectural design concept must be formulated in this phase (see Figure 5.2).

- **Purpose**
    - Ensure the project's functional requirements.
    - Ensure the project's technical requirements when this is important.
    - Plan the management of all knowledge areas.

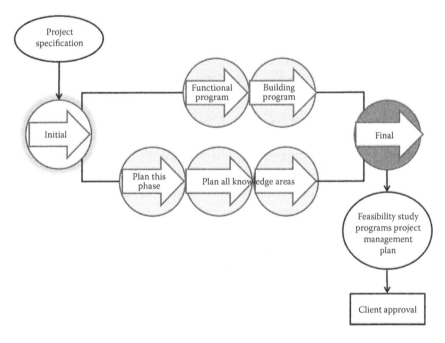

FIGURE 5.2
The process to get approval for the project management plan including the feasibility study and building programs.

- Determine how and when to do things and who is responsible for different parts of the project.

If the general planning procedure from Section 4.2 is transferred to an industrial and building project, the results of the work can contain the following headings:

Functional Program	Building Program	Management
Activity description	Site survey	Codes, permits, and approvals
Room-function program	Building location	Validation requirements
Environmental aspects	Existing building and land survey	Times
Safety, health, and environment	Building appearance program	Costs
Fire and explosion risks	Program sketches	Resources
Waste management	Structural principles	Responsibility matrix
Artistic program	Existing fittings survey	Communication plan

Operation and maintenance management	New machinery, equipment, fittings, and furnishings	Risk management plan
	Foundation work principles	Type of contract
	HVAC principles	Stakeholders' views
	Water supply and sewage principles	Project strategies
	Electrical principles	
	Fire and explosion rating plan	
	Flexibility about changes	
	Ground-level plan	

In this phase, it is necessary to determine type of contract and the desirable type of remuneration.

Type of reimbursement is discussed at this stage, but may be changed during the negotiations (see Section 6.12). Professional project management requires that all knowledge areas be planned, monitored, and controlled.

5.4 PROJECT IMPLEMENTATION AND DESIGN

The design phase describes the production of documents such as drawings and technical specifications. Consultants ensure quality and environmental control of their projects using ISO 9000 and ISO 14000.

The design work goes stepwise with successive approvals (toll gates; see Figure 5.3). Throughout the work, control can be exercised so that the documents will be ready in time, the production costs and production times are as planned, and the risks for the proposed solution are acceptable. Environmental requirements and the requirement for a safe work environment during production and for the finished product must be met. How uncertainties (threats and opportunities) for the production and final product affect the design must also be determined.

Coordination of work among the project teams—architects and construction, mechanical, HVAC, and electrical engineers—must start in this system/basic design phase (see Figure 5.4).

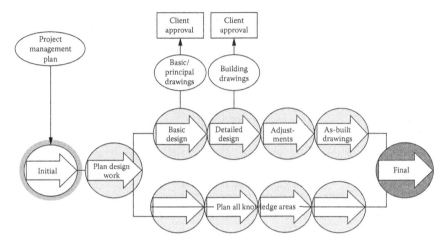

FIGURE 5.3
The flow of the design process with the project document should be approved.

5.4.1 System/Basic Design, Principal Drawings, Documents

- **Purpose**
  - Establish a technical and an aesthetic design.
  - Reconcile and summarize the situation in all knowledge areas.
  - Get basic documentation and approval to proceed with the project.

The system documents transfer functional requirements into technical documents. The solutions must be within the constraints of time, cost, risk strategy, and environment, health, and safety that are identified in the project management plan.

During this phase, meetings with users, facilities managers, operation and maintenance representatives, interior designers, and others are held. Within the company, the following stakeholders with special responsibility may need to be consulted:

- Persons responsible for contacts with authorities
- Fire questions: fire protection manager
- Security and camera surveillance: security chief
- Infectious disease issues, allergies, contamination (hospitals, research, etc.)
- Logistics
- Health and safety: safety manager

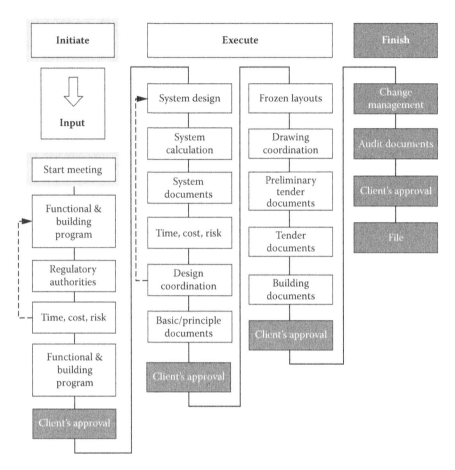

FIGURE 5.4
The design process.

- Environmental issues: environmental manager
- Telecommunications and data
- Security and automatic doors
- Validation: validation or quality manager

The project's aesthetic and architectural design must be approved and frozen at this point.

Scope and detail of the systems and principal documents depend on the type of project. At this stage, it is advisable to carry out principal layout plans for the water supply, sewage, electricity, and principal solutions in tight sections.

Principal drawings (PD) for electricity, water supply, sewage, and HVAC design must be prepared early in the design phase. The scope of these drawings is decided in the project manager's decision (PMD) meetings. The system and principal drawings for a project may include the following information:

- The main sections of the building
- Layouts for mechanical and HVAC rooms, substations, and electric and telecommunication centers
- Size of machinery and the need for temporary openings and equipment (scaffolding and platforms, both permanent and temporary)
- Principal flow charts for water supply, sewage, HVAC, and electrical installations
- Definitive location and size of larger openings
- Other space requirements
- Preliminary coordinated cross section of corridors with installations
- Preliminary coordinated cross section of apron under windows with installations
- Other agreed upon cross sections with demands of coordination
- Size and location of air intakes and discharges
- Preliminary elevator matrix (shaft and pit size, car size, lift and types of doors)
- Principles of fire subdivision, sprinkler, etc.
- Principal layouts and sections

An important component in this stage is to monitor and assess the project's cost at completion. If constraints are not met, it is necessary to redraw or ask for more money. At this stage not only needs, but also wishes, often show up. A strict handling of requests must be made to contain the budgetary and time constraints. Building permits are often sought during this phase.

System and principal documents can contain the following:

1. General
    1.1. Brief general description of the background and objectives
    1.2. Compilation of different area and volume data
    1.3. Compilation of various power and energy needs, water supply, sewage, and HVAC
    1.4. Compilation of various power and energy needs and electricity

1.5. Safety program
1.6. Summary of objective, end-effect goals, and functional and technical requirements
1.7. Site layout
2. Contacts with regulatory authorities
    2.1. Building codes and permissions and approvals needed
    2.2. Demands for validation
3. Decision, coordination, and integration
    3.1. Type of contract
    3.2. Interface with different disciplines
    3.3. Interface with other projects
    3.4. Interface with equipment, machinery, furnishings, etc.
4. Project properties
    4.1. Functional program (see Section 6.3.1)
        4.1.1. Brief description
        4.1.2. Good manufacturing practice, good laboratory practice, etc.
    4.2. Technical descriptions and drawings
        4.2.1. Room and materials descriptions (outline specification)
        4.2.2. Structural description
        4.2.3. HVAC water supply, sewage, and sprinkler description
        4.2.4. Master control center, devices, and electricity
        4.2.5. Description of electricity, telecom/data, and lift facilities
        4.2.6. Description of fire alarms, burglar alarms, and access control
        4.2.7. Description of landscaping and external water supply and sewage facilities
        4.2.8. Geotechnical investigation
        4.2.9. Description of fitments, fixtures, and furnishings
        4.2.10. Description of acoustic needs and solutions
        4.2.11. Description of fire protection
        4.2.12. Architectural and fire subdivision drawings
        4.2.13. Structural drawings
        4.2.14. Mechanical drawings
        4.2.15. HVAC drawings

4.2.16. Electrical, power, telecom/data, and elevator shaft/doors drawings

4.2.17. Drawings for landscaping and external water supply and sewage

4.2.18. Layouts and principal drawings of machinery, fitments, fixtures, and furnishings

4.2.19. Other drawings and descriptions

5. SHE (safety, health, and environment)
   5.1. Environment plan
   5.2. Work environment plan
6. Milestones and other critical dates
7. Economy
8. Special resources
9. Organization and staffing
10. Information and communication
    10.1. Distribution lists
11. Risk management plan
12. Procurement
    12.1. Long-lead-time items
    12.2. Special guarantee periods

5.4.2 Building Document Design and Preparation of Inquiry Documents for DBC Contracts

- **Purpose**
  - Create clear technical and administrative documents that will form the basis for inquiring, contracting, and production.
  - Through progressive design and coordination of drawings from different consultants, reduce redesign during the production phase and minimize the number of changes to contract documents.

5.4.2.1 Design Documents on the Way to Building Documents

Frozen basic layouts from the architect are made after consultation with users and aim to reduce the number of changes during further design work. The freeze means that users' activities, floor areas, floor heights, frame systems, load-bearing walls, pits, elevators, mechanical and HVAC shafts, and rooms for mechanical, HVAC, and transformer equipment are "frozen." The floor layout and corridors are resolved, but drywall, doors, etc. can still be changed by the user and architect. The frozen layout is the basis for the structural, mechanical, and electrical consultants when they start their drawings. Together with the users, the PM should check the layouts before they are frozen. Drawings and key technical specifications are signed by the parties involved. After this, modifications to the frozen layouts must be decided in special meetings. Changes by the user, architect, or structural or mechanical consultants can thus be controlled by the PM.

Frozen suspended ceilings are intended to provide installation consultants with enough time to plan all the installations in the often very limited space over the suspended ceiling when the space is tight or when there are hygienic demands. Is the suspended ceiling fixed or does it have panels? This information is very important. The PM should check with the users before the suspended ceilings are frozen. Modification of a frozen document may be made at decision meetings after consulting users and other consultants. In this way, changes can be controlled by the PM.

Halfway drawings give the consultants a comprehensive knowledge of the other consultants' flooring, windows, installation volumes, cable ducts, protruding beams, and insulation solutions. The PM can control whether the consultants are following the schedule and the solutions are within cost, risk acceptance, and an acceptable production duration.

As an aid in planning coordination, the project team can gather the following:

- Pit information (template; see Appendix C)
- Elevator information (template; see Appendix D)
- Door matrix (template; see Appendix E)

The architect's halfway drawings should contain at least the following information:

- Fire subdivisions
- Layouts and principal drawings of machinery, fitments, fixtures, and furnishings
- Facades, including location of rainwater pipes
- Main sections
- Principal vertical sections, for example:
  - Preliminary coordinated cross section of corridors with many installations

- Preliminary coordinated cross section of apron under window with many installations
- Interface façade/terrain
- Eaves and mirrored eaves
- Air intakes and exhausts
- Suspended ceilings, including connection to the facade
- Complicated constructive details
- Thermal convection details
- Wall elevations with principal horizontal sections showing:
  - Windows
  - Top layer of stone
  - Top layer of wood, etc.
- Floor plans showing surfaces of:
  - Stone or terrazzo
  - Plastic mats
  - Linoleum
  - Epoxy
  - Wood flooring
  - Other areas that have an influence on structural drawings
- Plans and sections showing floor openings for installations
- Location of fire hydrants and emergency showers and eye washes
- Stairs
- Contract limits (if required)
- Door matrix

The structural engineer's halfway drawings should contain at least the following information:

- Excavation plan with location of sheet pile walls
- Preliminary pile situation plan for coordination with below-floor slab installations
- Water supply and sewage works below the floor slab; these drawings can also be made by the mechanical engineers
- Floor slabs, not yet including reinforcement
- Main sections showing the location and dimensions of columns and beams
- Plans and sections showing floor openings for installations
- Report of self-oscillation beams and slabs
- Elevation of concrete walls showing holes

- Plans and elevations of the steel frame
- Coefficient of thermal transmitters for exterior walls and windows
- Chimney mountings
- Rafters
- Principal vertical section showing, for example,
  - Eaves
  - Interface facades and terrain
  - Complex constructive sections
- Principal components showing, for example,
  - Thermal bridge problem
  - Sealing problems
- Ground plans showing new and existing pipes and cables (together with other consultants)

The HVAC engineer's and mechanical engineer's halfway drawings should contain at least the following information:

- Plans showing the locations and preliminary design of large ventilation arrangements, ducts, etc.
- Layout for air handling and plant rooms
- The main sections showing locations and preliminary dimensions of piping and HVAC
- Plans and sections showing the use of floor openings for installations
- Principles of fire insulation and sectioning
- Floor plans showing new and existing pipes under the floor and ground plans showing external new and existing pipes (together with other consultants)
- List of needed holes in concrete walls for pipes and ducts
- Locations of convectors and radiators, and space requirements
- Installation above suspended ceilings
- Layout for sprinkler
- Locations and sizes of floor drains and gutters
- HVAC load calculation showing different alternatives for heating, cooling, and ventilation
- Locations and loads of large tanks
- Need for lifting beams (operational needs)
- Needed openings in steel and concrete structures for mechanical installations

- Required openings and loads (temporary) for transportation of large equipment into the building

The electrical engineer's halfway drawings should contain at least the following information:

- Principal outlet locations
- Principal fixture locations
- Main sections
- Principal plan for light fittings
- Principal plan for power supply and main cable layout
- Principal plan for power, telecom/data, and HVAC control
- Layout for cable runs and power, telecom/data, HVAC control, security cables
- Principal plan for standby power
- Principal sections showing cable runs
- Principal fire alarm system
- Needed openings in concrete and steel constructions for electrical installations
- Information on elevator type, trench, pit size, basket size, door types, machine room, and doors
- Fire and explosion classification plan
- Required openings and loads (temporary) for transportation of large equipment into the building

The halfway drawings for machinery, fitments, fixtures, and furnishings should contain at least the following information:

- Service requirement and influence on power, water, sewage, air, air change rate, extracted air, and heat loads to be taken care of
- Principal coordination of mechanical, HVAC, electrical, machinery, fitments, fixtures, and furnishings

P-drawings are preliminary inquiry documents, where only one or two descriptive texts and certain dimensions may be missing. These drawings will give the consultants a final opportunity to coordinate their documents. In addition to drawings, the documents should include technical descriptions, administrative regulations, the site plan, and an inquiry schedule. All project participants should review the P-drawings.

Quality surveyors and building, mechanical, and electrical inspectors on the site should review and propose improvements to or clarifications of the P-drawings. These changes should be included in the inquiry documents if the PM and the consultant can take responsibility for them.

Electrical, mechanical, and HVAC engineers should establish **building PMs.** These will include information about products and work that normally are carried out by the building contractor in the interface with installation work—for example, foundations and steel frames for heavy machinery, fan guards on roofs, slits, holes, inspection holes, steps, and walkways for the servicing of the HVAC plant. Details and work descriptions of these building PMs should be incorporated into the architectural and structural engineering documents.

Three-, four-, and five-dimensional (3D, 4D, 5D) building information management (BIM) documents will help owners, users, consultants, and contractors to analyze 2D drawings. They also offer collision control for the building and installation design. In many countries, it is not yet clear who owns the model and this must be cleared up by the PM before the BIM design starts.

In 2011, 3D design programs that promised to detect collisions did not always deliver what they promised. Personal experience has shown that details, such as reinforcements around holes in steel sheets, are not always checked with the installations.

The inquiry document contains drawings, specifications, and contract administrative documents that are the base for bids. **Construction/building documents** are used for production. The agreements made during the negotiations must be incorporated into the building documents. Unfortunately, construction/building documents are often changed during production because of change requests from clients or users, missing information, or poor coordination. **Labeled information of electrical and plumbing installations** is often forgotten in the inquiry documents. A special supplement should be prepared and included with the inquiry documents.

Holes in walls and floors often create discussion. For turnkey (TK), design–construct (DC), and general construction (GC) contracts, this is not the PM's headache; however, for coordinated general construction (CGC) contracts, it is important that the principles for making holes for other contractors are clear. **Tolerances** are an important parameter in designing and building. A business-like building contract is not possible if millimeter tolerances are requested. Smaller tolerances result in higher costs. A skilled architect and structural engineer can hide the tolerance requirements. If not, the building will have a sloppy appearance. **Choice of color scheme** is often missing in inquiry documents. A color scheme that is prepared after the contract is awarded can result in unnecessary discussion about additional economic compensation.

5.4.3 Detailed Design: Inquiry and Building Documents for DC/TK Contracts

5.4.3.1 Inquiry Documents

- **Purpose**
  - Ensure overall objectives, without limiting the possibility for the DC/TK contractor to produce economic solutions.

In a TK/DC enquiry, the feasibility study and programming (see Section 5.3) are supplemented with geotechnical investigation, an outline overall plan, a possible outline layout showing the requirements of the functional program, and a concept of aesthetic design (see Figure 5.5). Inquiry documents for equipment are supplemented with a room layout, permitted transportation routes, and load and space constraints.

Among other things, the design program and outline specification drawn by the architect should contain material requirements, surface treatment requirements, and special requirements on hygiene, cleaning, and maintenance.

For electrical, mechanical, and HVAC contracts, the design program and outline specifications should describe the functional requirements, technical principles, materials, grades, and financial objectives relating to energy efficiency and environmental demands.

All programs and outlines of overall plans should include objectives relating to operation and maintenance. A good PM will discuss these issues with the operation and maintenance department. Any limitations on product selection should be documented. The DC contractor's total production of documents follows the process flow for the design of DBC (design–bid–construct) contracts (see Section 5.4.2).

5.4.3.2 Production of Building Documents in DC Contracts

- **Purpose**

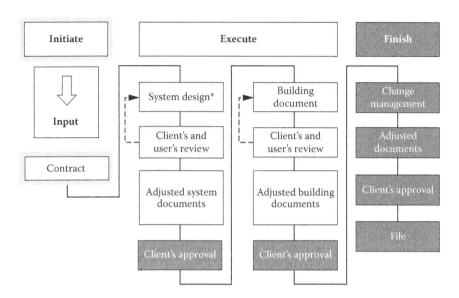

FIGURE 5.5
The design process for DC and TK projects. The input is the contract documents.

- Ensure that proposed solutions will be what the customer intended when purchasing and, if necessary, that changes are ordered before production begins.

To ensure that the drawn product conforms to the contract requirements, the client lets the design inspectors continuously review the design work. The consultants who prepare the programs and outline specifications might be an appropriate choice. However, it is necessary to keep in mind potential conflicts of interest.

5.4.3.3 Preparation of Documents for Equipment, Fitments, Fixtures, and Furnishings

- **Purpose**
  - Ensure that proposed solutions will be what the customer intended when purchasing and, if necessary, that changes are ordered before production begins.

In order to coordinate equipment and furnishings with other planning, it is important to obtain and share, as early as possible, information about

- Energy and heat emission
- Media connections, volume, connection dimensions, and specifications
- Constraints and interfaces between project parts: What information should be provided? Who should inform whom? When must the exchange take place?
- Requirements related to the factory acceptance test, site acceptance tests, and coordinated testing

In complex projects, it may be useful to have an interface information process that clearly shows the when, what, where, and how of information and installation. (See Appendices F, G, and H.)

5.4.3.4 As-Built Documentation

- **Purpose**
  - Ensure that true information on buildings, installations and equipment, fitments, fixtures, and furnishings is available.

As-built documentation (drawings, descriptions, and specifications) should be prepared and filed according to the client's routines and specifications before the project is taken over by the client. Sometimes as-built documentation is included in the contract. At other times, the contractor submits information to the client and its consultants and the operational department is responsible for the preparation of as-built documents. In the latter case, the client must arrange and transfer responsibility for the as-built documentation.

If the as-built documents are computer-assisted design (CAD) documents, the client's CAD standards must be included in the contract documents.

5.4.3.5 As-Built Information and Documentation from Contractors and Suppliers

The client's inspectors will review and transmit approved information and documents to the consultants for them to do as-built documents (see Figure 5.6). This is appropriate with written confirmation in connection with the delivery of information documents between the contractor, inspector, consultant, and PM. Archiving documents is described in Section 6.10.6.

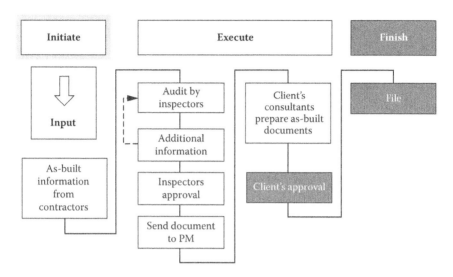

FIGURE 5.6
Process for producing as-built documentation.

5.5 PRODUCTION AND INSTALLATION PHASE

Many contractors and suppliers have quality and environmental management plans based on ISO 9000 and ISO 14000. The task leader (TL) and contracts manager (CM) must also manage their tasks professionally. They must study all phases and knowledge areas in the project management process. The TL and CM must prepare a project management plan for "their" project and prepare reference plans to be able to monitor their work. This is done to work and control the processes in the different knowledge areas (see Figure 5.7).

The contractor often becomes the client of subcontractors and material suppliers and must make it clear what applies in the procurement knowledge area.

5.6 GUARANTEE PHASE

The project's product—a building, a road, a railway, etc.—is often transferred from the PM to the operations and maintenance department. The knowledge of the project must be transferred correctly to avoid valuable

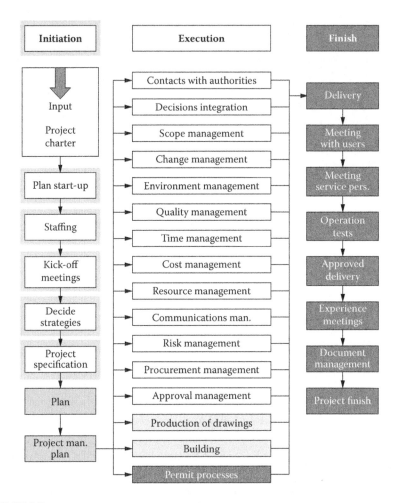

FIGURE 5.7

Project processes with illustration of the project management knowledge areas.

waste. This is normally done through education and delivery of as-built drawings and operation and maintenance instructions. The handover can be very smooth if the operation and maintenance personnel are involved in testing, inspections, and coordinated tests.

Traditionally, the mechanical, HVAC, and electrical contractors transfer operating and maintenance instructions and lists of installed products. The corresponding operating instructions and lists from the building contractors regarding materials (e.g., cleaning instructions for plastic mats and product specifications for door closers and ceilings) that are needed for the product's operation and maintenance should also be transferred.

During the warranty period, noted actions, errors, and deficiencies should be reported to the guarantee inspector who, at the guarantee inspection, must determine whether there is a warranty issue or not. Before the guarantee period is closed, an inspection meeting should be held (see Section 6.2.2). The information from this meeting will be brought up at the guarantee inspection.

The guarantee inspection must be held before the guarantees expire. Note that there may be multiple guarantee periods for the same building. The guarantee period may also vary for different products and work. Special guarantees apply to "latent defects." The inspector can sometimes extend the guarantee period for specific items or work.

During the project's final phase, the person responsible for contacting the guarantee inspector and the person who should contact all parties involved in these guarantee inspections must be determined. When the defects are solved and approved, the client should return all bonds and bank guarantees or equivalent documents.

6

Knowledge Areas

The text of this chapter is based on a hypothetical industrial facility with offices, research and production facilities, and central media supply with high standards of delivery performance. This can be used in the preparation of a project management plan. Those named as responsible are suggestions—not requirements.

6.1 CONTACTS WITH AUTHORITIES

WHAT permits and approvals are needed?
HOW and **WHEN** can these permits be obtained?

The regulatory process is generally as shown in Figure 6.1.

6.1.1 Permits for Building, Demolition, Earthwork, and Temporary Measures

In order to protect citizens, third parties, and other interests related to construction projects, a number of federal, state, and local laws, ordinances, and regulations must be followed. Building codes provide minimum standards for the protection of life, limb, property, and environment and for the safety and welfare of the consumer, general public, and the owners and occupants of residential buildings regulated by this code (see Section 2.9 in Chapter 2).

Codes are local business. Always check what is valid in a particular area!

Before a project starts, permits are necessary. Keep in mind that the authorities' handling time can often be long. Permission should be granted prior to procurement of services. In other cases, responsibility

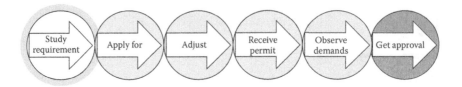

FIGURE 6.1
The contract process flow.

for the project permit should be passed on to the main contractor, who assumes responsibility for the process of developing the building permit documents (see Figure 6.2).

A **building permit** is the first step in the construction process. The permit is needed to erect a new building or structure (it can include fences, rock walls, and animal shelters over a certain height; towers; and retaining walls). It may be needed to install energy-saving devices (solar systems, wind towers, etc.). It is needed to relocate, demolish, repair, alter, change the occupancy of, or make additions to an existing building or structure.

A **demolition permit** authorizes the removal of an existing building or a part of it. An authority may issue a demolition permit for a residential or nonresidential structure. Sometimes permits are required for interior remodeling for the removal of plaster, lath, and drywall. There can be problems obtaining a demolition permit for buildings of historic, cultural, or other interest.

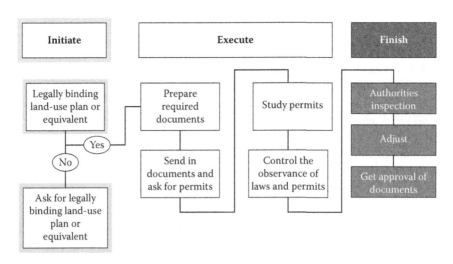

FIGURE 6.2
The process of contacts with authorities.

There may be a call for a **demolition plan** as a basis for sorting and recycling of demolition materials. The local authority decides on the level of demolition plan.

An on-site **sewage disposal permit** is needed to secure sewage treatment and disposal for homes not served by a community sewer system. Drain fields must be properly designed and installed for two reasons:

- To protect the ground water aquifer from sewage contamination
- To protect the health of the populace by properly treating and disposing of human waste

An **earthwork permit** is needed to preserve the natural environment and the stability of hillsides and slopes, and to control erosion and sedimentation by the regulation of earth-disturbing activities, excavating, and filling of land. An **earth movement permit** may also be needed. It is necessary to consider

- Keeping downstream flooding, erosion, and sedimentation at existing levels
- Reducing damage to receiving streams and impairment of their capacity, which may be caused by sedimentation
- Protecting the stability of sensitive slopes

If work is to be performed in or near wetlands, a **wetlands permit** is needed. Some wetland types are common, while others are rare. All, however, provide some valuable functions and resources. These may be ecological, economic, recreational, or aesthetic.

Federal, state (in the United States), and local governments may all have specific permit requirements. At the federal level, the US Army Corps of Engineers regulates wetlands under the Clean Water Act and US Coastal Zone Management Act.

Activities That Need Permits

Activity	Example
Deposit fill material	Bulldozing, grading, dumping
Dredge or remove soil or minerals	Removing tree stumps, bulldozing, digging a pond
Construct, operate, or maintain any development	Construction of buildings or structures; boardwalks; peat mining; water treatment
Drain surface water	Diverting water to another area via ditch, pump, or drain

In some countries, it is necessary to include **wartime shelters** in a project. The requirements for shelters vary. Sometimes there are central shelters, so there will not be any shelter claims for the project. The defense and/or civil defense areas can sometimes have views on these shelters.

Designate responsiblity so that

- The requirements for permits and approvals are developed early in the project
- Applications for permits and approvals are submitted sufficiently ahead of time
- The demolition plan is prepared
- The issue of shelters is clarified
- A fire description is drawn up and the fire protection documentation is ready for final inspection (when required)

6.1.2 Quality Manager for Approval of Permits

Many countries require that a certified person review the project and declare in writing that the project fulfills all requirements concerning codes and permits. In other countries, the authority has its own inspectors review the whole project. This quality function is not the same as that for project quality management. The quality manager (QM) for permits only controls what codes and permits to request. The project quality manager (PQM) must control all quality issues for the project.

6.1.3 Approvals of Elevators, Pressure Vessels, Refrigeration Plants, Transformers, and Other Electrical Plants

In most countries, some items, such as elevators, machines, autoclaves, pressure vessels, transformers, electrical plants, etc., must be approved by the authorities or specially authorized companies. The responsibility for these inspections should be on the contractor. Remember to include this in the contract. A client who is responsible for these inspections must be included in the planning. **What** is needed to conduct the test? **When** should the tests be done?

For elevator inspection, who is required to supply the loads in the elevator car? It is easiest if the elevator contractor provides this.

Designate responsibility so that

- The surveys that do not lie with the contractors/suppliers are carried out
- The authorized inspection of commercially available equipment is carried out by the supplier or by the client

In the United States, a number of independent laboratories perform tests of products for the US market. The Occupational Safety and Health Administration (OSHA) maintains a list of approved testing laboratories, known as nationally recognized testing laboratories.

6.1.4 CE Marking, the EMC Requirement, and Testing

In European countries, before a machine or plant is taken into operation, it must be CE (abbreviation of *conformité Européenne,* meaning European conformity) marked. The **CE marking** on a product is a manufacturer's declaration that the product complies with the **essential requirements** of the relevant European health, safety, and environmental protection legislation. CE marking on a product indicates to government officials that the product may be **legally placed on the market** in their country.

A product's **electromagnetic compatibility, EMC,** means that the electromagnetic disturbance it generates does not exceed a level allowing radio and telecommunication equipment and other apparatuses to operate as intended, and that the apparatus has an adequate level of intrinsic immunity to electromagnetic disturbance to enable it to operate as intended. The directive applies to "all electrical and electronic appliances" together with equipment and installations containing electrical and/or electronic components liable to cause electromagnetic disturbance, the performance of which is liable to be affected by such disturbance.

6.1.5 Fire Prevention and Fire Safety Documentation

The Engineering Services Section of the Minnesota Department of Health (2011) has the following text on the Internet concerning fire prevention and fire safety documentation:

> NFPA 101(00) and other codes require that a fire safety and evacuation plan is prepared and maintained for healthcare occupancies. The codes expect that all employees will receive on-going training with respect to their duties under this plan. This is typically done through annual staff training and periodic fire drills. In addition, a plan must be in place detailing how

a facility will handle situations in which the building fire sprinkler system and/or fire alarm system are out of service.

The codes also contain requirements relating to the flame resistance of drapes, curtains and decorations, the flame spread rating of interior finishes and, by reference to other NFPA standards, the testing, inspection and maintenance of fire protection systems.

Here, requirements for fire protection, safety and evacuation plans, and descriptions of how fire protection with sprinklers, etc., works can be found. Most countries have similar documentation requirements.

Remember! Codes are local business. Always check what is valid in the particular area of work!

6.2 DECISIONS, COORDINATION, AND INTEGRATION

WHAT are the powers of different project participants? (See Section 4.1.1 in Chapter 4.)

WHAT decisions are needed to implement the project?

WHAT is necessary to coordinate within the project and **WHAT** user activities are affected by the project?

WHAT is necessary to know about the project's gradual integration with the client's existing space and facilities?

WHAT is necessary to know about the company's gradual takeover of the project?

WHAT is necessary to know about the company's operating capacity related to water, heat, and electricity?

WHAT capacity of the preceding services can be used for the project?

WHAT are the crucial decisions? **HOW** and **WHEN** are these decisions obtained from client, users, designers, contractors, and suppliers?

HOW and **WHEN** is it determined and documented how the coordination should be handled in order to minimize ambiguous acts and disruption of users' ongoing activities?

HOW and **WHEN** are the designers' and contractors' ongoing businesses integrated with the project activities?

HOW and **WHEN** are the users' and client's activities integrated with the project?

HOW and **WHEN** is access to the surfaces and premises obtained?

HOW and **WHEN** can existing media be used for temporary water, heat, and electricity?

6.2.1 General

Control is done by monitoring and acting when something does not follow the plan. If the various knowledge areas, processes, and developments are monitored and changes made as needed, the project work can be well controlled.

Project managers (PMs) control the project through decisions at meetings and through written or oral directives. Coordination must be done between the project and the environment where the project takes place. When facilities are renovated or added, a project must be coordinated with current activities in existing operations. The project's need of designers, contractors, and suppliers must also be coordinated with other projects. Transport, electricity and water suspension, etc., must be coordinated with operations and possibly third parties in the surrounding area. This is often done at meetings.

Gathering of requirements and information, information distribution, choices between alternatives, project coordination, interface management with the "world outside the project," and decisions on changes are often made with the help of meetings. A structured meeting culture is important. The easiest way is to divide the meetings into decision meetings, coordination meetings, infrastructure meetings, briefings, and meetings to resolve technical details. Sometimes, two consultants in a decision meeting may suddenly begin to discuss technical coordination while five or six people are sitting idle. Wasted time! This does not mean that several meetings should be held in succession with the same people attending. However, remember to keep the issues apart.

6.2.2 General Meetings

Meetings should be kept short and participants should be well prepared. Meetings and meeting rooms should be booked well in advance. Agendas should be attached to the meeting summons. Failure to do so means that participants cannot prepare their questions. It should be clear which persons are responsible for information on each point. It can be convenient to have previous protocols as the agenda, maybe supplemented with new questions. In order to have time to study the new agenda, all meeting participants should be notified at least 3 days before the meeting.

Minutes from meetings are working tools and should be available to participants within 3 working days after the meeting. The documentation is easier if there are templates and examples of invitations, agendas, and minutes from meetings.

- **Roles and responsibilities**
  - Designate
    - **Who** has to send out invitations to meeting?
    - **Who** is responsible for booking a conference room?**Who** prepares the agenda?
    - **Who** will chair the meeting?
    - **Who** writes protocols?
  - Meeting climate: If the meeting climate is bad, there can be no good decision. To have good meetings, follow these general rules:
    - Stop digging. Do not look back and find fault. Look ahead and make decisions about what must be done.
    - Be sensitive and respect others' opinions.
    - Avoid personal attacks.
    - Do not interrupt. Speak clearly and briefly.
    - Do not evaluate because this will interrupt the speaker.
    - Avoid external disturbances, such as mobile phones, secretaries' interruptions, etc.
    - Arrive in good time out of respect for other participants and their time.
  - Respect what is confidential.

Sometimes, at meetings, things are discussed for a seeming eternity so that no one is disregarded and/or slighted. The chairman of the meeting must clearly indicate what the decision is and who has the delegated responsibility. This is especially important in international projects. Individuals may come from different cultures and expect crystal clear decisions and not a shared sense of what is decided.

Make sure that the participants of decision meetings are empowered to make decisions. A real anticlimax occurs when, after long discussions, the group is prepared to take a decision, but someone says, "I'm taking this information with me…" What kind of decision meeting is that?

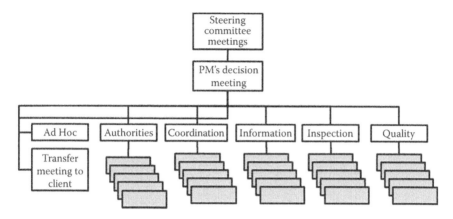

FIGURE 6.3
Structure for project management meetings.

6.2.2.1 Examples of Meetings in an Industrial Project

A PMD meeting is a project manager's decision meeting. An ad hoc meeting is a meeting that takes place when there suddenly is a need to discuss a specific purpose or event.

Figure 6.3 shows an example of a client's project management meeting that may be needed for complex industrial projects. For simpler projects, the number of meetings is reduced, but the list shows information that should be collected and discussed.

The following table gives examples and brief descriptions of the aims of various meetings and, sometimes, suggestions of participants. The design group and the general contractor/turnkey contractor establish appropriate corresponding lists concerning their design and production.

Meeting Category	Type of Meeting	Comment
Decision meeting	Steering committee meeting	
	PMD meeting (project manager's decision meeting)	
	Site meetings with contractors	
	Designated decision meeting, if necessary (ad hoc)	
	Transfer meeting to client or users	
Authority meeting	Information meetings with the fire and emergency departments	

Meeting Category	Type of Meeting	Comment
	Meeting with the area's environment and health departments	Can apply to kitchens, restaurant, food processing, etc.
	Working-environment safety and health department	
	Authority for special activities	Radiation, FDA, air fields, electromagnetic compatibility requirements, wetlands, etc.
Information meeting	Kick-off meeting	
	Exchange of experience from earlier projects	
	Program meeting to inform on objectives and strategies	
	Fire protection meeting	
	Safety meeting	
	S-meeting	
	Telecom/IT meeting	
	Workers' protection	
	Key system	
	Follow-up meeting with users and operation or maintenance personnel	
	Exchange of experience from designers and contractors	
	Closing meeting with turnkey, general, and subcontractors	
	Guarantee meeting with users and operators	
Coordination meeting	DC meeting (designer coordination meetings)	All consultants take their own notes
	CAD coordination meeting	
	Risk analysis meeting for the project	
	Risk analysis meeting for design	
	Start-up meeting with GE (general contractor)	
	Risk analysis meeting for production	
	Subcontractor meetings	
	Health and safety coordination meeting	

Meeting Category	Type of Meeting	Comment
	Equipment installation meeting	
	Environment coordination meeting	
Inspections	Standard or normative inspections	
	FAT (factory acceptance test)	
	Inspection and tests of elevators, pressure vessels, etc.	
	Fire protection inspections	
	SAT (site acceptance test) 1, SAT 2, SAT 3	
	Test on completion	
	Warranty and guarantee tests and inspections	
Quality meeting	Technology meeting	
	Production meeting	
	CE meeting	
	EMC meeting	
	Preparatory quality meeting	Plan quality work and explain the purpose of the project's quality meetings

6.2.2.2 Meetings with Authorities

- **Purpose**
  - Clarify requirements, processing times, and decision dates.
  - Get feedback on the aesthetic and technical proposals.
  - Provide information on environmental and safety issues.
  - Get approvals.

Meetings with government agencies will be held in connection with building permits and fire safety issues at the start of construction and the finished project. Meetings with the authorities should always be recorded by one of the project participants. A copy of notes from the meeting will be distributed to the appropriate authority. The project manager must approve these notes before they are distributed. They should be distributed to participants in the project as an appendix to PMD minutes.

Before a meeting with an authority, the PM must be informed of the reason for the meeting and the issues to be addressed. The PM will determine which persons should attend.

- **Responsibility**
  - The PM decides who attends meetings with authorities and who is the client's representative.
  - The client's representative creates "notes from..." and is responsible for their distribution.

An information meeting with the **rescue service** is sometimes held to ensure that proposed solutions will be accepted for a building permit application. Information meetings with the **fire inspection department** will be held during the planning phase to inform and ensure that proposed solutions will be accepted for a building permit application. Inspections by the **fire inspection department** should be done to have the project approved in connection with the tests on completion.

Meetings and inspections with **special authorities** concerning radiation, the Federal Department of Agriculture (FDA), air fields, electromagnetic compatibility requirements, board of agriculture, wetlands, or other authorities should be held to clarify requirements and determine whether special permits are required.

6.2.2.3 *Meetings during the Design Phase*

- **Purpose**
  - Provide information about changes and schedule review.
  - Decide on technical solutions and designs.
  - Coordinate technical solutions.
  - Obtain information from users and others.

The purpose of the **project start-up meeting** is to give everyone in the organization as much general information as possible to avoid confusion and facilitate administrative procedures. At this meeting, the project is presented to the project participants. The PM describes the proposed organization for the project and the planned type of construction and an outline schedule. Participants will also be informed about the procedures that apply regarding:

• Organization	• Documentation	• Quality
• Communications	• Meetings	• Contacts with authorities
• Health and safety policy	• Environmental policy	• CAD coordination
• Coordination reviews	• Approvals of documents	• Invoice routines

CAD (computer-assisted design) coordination meetings must be held to inform about and coordinate CAD work on the project. Decisions concerning deviations from the CAD manual may not be taken at these meetings, but rather submitted to PMD meetings for a decision.

MIS (maintenance information system) meetings are held to compile operation and management information, which is often overlooked. The information structure must have been discussed during the planning and the decision made as to how to inform and distribute. Separate MIS meetings can be held to inform and coordinate the work during both the design and production stages.

Risk analysis meetings for the whole project, for the planning phase, and for the production phase must be implemented (see Section 6.11.4).

Experience meetings should be held during the system/basic design phase to discuss lessons learned from previous similar projects. The PM, users, task leader (TL), contracts manager (CM), designs manager (DM), inspectors, operations/maintenance staff, and consultants will attend the meeting. If possible, participants should visit existing premises with the same or similar activities.

At the **steering committee meeting,** the comprehensive decisions are taken. The client or customer, the project manager, and executives of various businesses attend. In order not to lose time, the change decision of the project framework must be documented in a written protocol. These should promptly (within a day or two of the decision) be distributed to project participants.

PMD meetings are the PM's decision meetings during the planning phase. These meetings are normally held every 14 days and are chaired by the PM. All PMD meetings must be recorded. All decisions at PMD meetings should immediately be checked by the designers and project participants; if any doubts are raised, this should be discussed during the first DC meeting after the protocol is available. Any ambiguities or undocumented changes should be brought to attention at the subsequent PMD meeting for clarification. Information from **information meetings**, **meetings with authorities,** and **DC meetings** should be reported in PMD meetings. Minutes from the PMD meetings should be verified at the next meeting.

DC (designer coordination) meetings are the design team's coordination meetings during the planning phase. These meetings are normally held every 2 weeks and conducted by the DM. All consultants are responsible for obtaining the necessary information and documenting this themselves. Both the person sending and the one receiving information must act so that necessary information is delivered on time from user representatives and outside consultants. If late exchange of information will delay the design work, it must be discussed at the DC and PMD meetings. Sent and received information should be documented by the project participants and coordinated before the inquiry documents are completed.

Program meetings are held when it is necessary to clarify the users' requirements and to present solutions for the users. Short minutes from meetings are written. These minutes should be attached to the PM meeting minutes. The PM must examine whether the requirements at these meetings are needs or wishes and must be careful addressing late wishes (see Section 3.3 in Chapter 3).

Fire protection meetings will be held before the system/basic design drawings are approved by the users and before the bid request documents are sent out. At these meetings, all questions regarding the handling of flammable and explosive products must be clarified. The solutions to fulfill the fire protection regulations must also be clarified. Minutes should be written and should be attached to minutes from the PMD meetings. Any decisions at these meetings must be confirmed at the PMD meeting.

The meetings should address the following issues:

• Fire zone subdivision	• Automatic fire alarm installation
• Fire resistance class of the building	• Access for emergency services
• Classification plan	• Evacuation issues
• Sprinkler installation	• Ventilation of fire gas
• Indoor and outdoor fire hydrants	• Emergency signs
• Locations and types of fire extinguishers	• Documentation of fire protection system
• Filing door requirements	• Flammable goods

Safety meetings will be held to clarify the safety, surveillance, escape routes, accessibility, and parking for final product during production time.

S-meetings (safety, low voltage, door control, surveillance camera meetings) should be held before the system/basic design drawings are approved by the users and before the request documents are sent out. This type of meeting is held because coordination and assembly problems often occur during the production phase because the program was not clear enough. Minutes of the meeting should be attached to PMD meeting minutes. Proposals from this meeting should be reviewed by the PM, so it is important to write clear and explanatory protocols. The meeting will address the following topics:

• Entrance and access control system	• Burglar alarm installation
• Electrical lock installations	• Automatic door openers
• Individual solo alarms	• Boundary alarms
• Alarms from elevators	• Alarms from restrooms and toilets
• CCTV surveillance	• Emergency alarm
• Location of door switches for handicapped persons	• Door matrix for door production (see Appendix E)

Telephone and IT meetings should be held before the schematic/basic design drawings are approved by the users and before the bid request documents are sent out. Minutes should be written and attached to PMD meeting minutes. Proposals from these meetings are decided on at the PMD meetings. The meeting will address and clarify the following points:

• National telephone facilities	• Any direct telephone routes
• Elevator telephone facilities	• International television
• Rooms for incoming data	• Cable TV networks
• Telecom/IT niches and communication room	• Central clock system
• CCTV surveillance	• Telephone switch room
• Entry phones	• Cable layout

Work/safety meetings should be held before the system/basic design drawings are approved by the users and before the request documents are sent out. Minutes should be written from the meeting and attached to PMD meeting minutes. Proposals from these meetings are decided at the PMD meetings. These meetings should follow the requirements in the Occupational Safety and Health Act to assure safe and healthful working conditions for working men and women.

Meetings on door fittings and locks should be held in good time before request documents are sent out. Representatives from the company's security department should attend the meeting. For security reasons, the lock matrix is distributed to very few people. The architect usually describes the door lock fittings and the operating/security department obtains and installs cylinders.

The door lock fittings will be delivered and installed by the building contractor. The architect must also describe the need for cable ducts in their frames and blades for automatic doors, entry systems, etc. (Compare to the S-meetings.) The information is assembled in a door matrix (see Appendix E). Suggestions from this meeting are decided on in PMD meetings.

The client's installation of cylinders should be coordinated with the takeover of the facility (final inspection) as the responsibility for the facility often is transferred from the contractor to the client at this occasion. For completion of outstanding work, especially in apartments, it is important that the key issue is handled seriously and checked. The architect is normally responsible for making sure that the meetings are held well before the requirement documents are completed.

6.2.2.4 Meetings during the Production Phase

- **Purpose**
  - Provide information about change management.
  - Monitor production status and small indications of any time extensions.
  - Clarify any requirements of reimbursement due to contract changes.
  - Facilitate production and assure performance quality by explaining or justifying why certain technologies have been selected.
  - Coordinate production within contract and with outside contractors.
  - Get feedback on alternative profitable production designs.
  - Settle time and cost claims due to contract changes.

Kick-off meeting are held with contractors and outside contractors to discuss:

• Site layout; huts, containers, storage areas; free transportation roads for client's users and parking for contractors	• OSHA; safety coordination meetings; safety inspections
• Contacts with authorities; building permit	• Bonds; insurance
• Perimeter; badges	• Payment plan
• Site and building boundaries; alarms and entrance security cards	• Quality plans
• Procedures when accidents or fire occurs	• Environmental plans and issues
• Water, power shutoff	• Contract review (special points)

Site meetings are normally held with contractors every 14 days and with subcontractors (SCs) when needed. The supervisor for the big SC should attend. The site manager (SM) may require that the supervisor of a specific

SC be present at meetings to discuss specific issues. The supervisors of fittings and furniture contracts, etc., can, when necessary, also be called to the meeting. Minutes from site meetings are normally confirmed at the next meeting.

SC meetings are held by the main contractor. The contract should allow the HVAC, mechanical, and electrical inspectors to attend these meetings to be informed and comment on the performance.

Client operations information meetings are held to inform external personnel about the client's production and activities in order to create an understanding for the special security and operational conditions.

Review meetings with legal representatives (LRs). In large companies with many projects, the legal representatives cannot attend all meetings. The time for reading minutes and attending information sessions is often limited. To inform LRs about the project development and discuss possible improvements, review meetings can be held. Small disputes can be solved at these meetings before they turn into conflicts and destroy an ongoing collaboration.

Risk analysis meetings for the production should be conducted by the SM and the contractors, inspectors, and sometimes a designer. The uncertainties and risks (see Section 6.11.4) should be discussed at these meetings.

Technical information meetings (kick-off meeting) should be held for those parts of the work that the CM or inspectors believe can create production or quality problems. The consultants' technical solutions should be clarified for the installations manager (IM) and, above all, for the foreman who will lead the job. Comments from supervisors and foremen will be considered by the PM and designers. Examples of technology information meetings' subjects include complicated sheet-metal work, sheet piling, visible moldings, plaster work, gas piping, air handling, sealing around pipes, and special painting.

Production technology meetings can be held early to identify specific production difficulties such as critical passages or areas with many installations by different contractors at the same time.

Health and safety meetings are held to coordinate protection at the workplace. These meetings are supplemented by safety inspections. The meetings should inform about current fire safety issues, whether alterations or extensions are needed, and who from the client is responsible for coordinating this with the client's operations. The SM normally calls for the first meeting. The OSHA instruction will decide how to continue.

Preparatory quality meetings should be held by the client's quality representative, contractors' quality managers, and inspectors to discuss how quality work will be conducted in the project.

Quality meetings are held regularly between the SM, client's quality representative, contractors' quality managers, and inspectors. At these meetings, the **construction company's quality manager** should verify how the quality work is performed in this particular project. The question should not be delegated to the quality manager at the site. It is the contractor or supplier who should demonstrate that the quality delivered meets the contract requirements. The client's inspectors should not have to point out where the quality requirements are not met. For example, in the automobile industry, it is not car buyers who should discover the errors; rather, the car manufacturer should ensure quality compliance.

In connection with these meetings, inspections on the site are performed and signed checklists and control plans are reviewed. If there are disagreements between the SM and the contractor at this stage, the final work inspector can be called in to review the required contract standard. The contractor should add to the client's report his or her own defect reports, which can affect the final product or where a temporary activity requires the client's approval.

The outcome of these meetings should be reported to the PM, who can take action, such as the deferral of payments, if the quality work does not take place.

6.2.2.5 Meetings during the Project's Final Phase

- **Purpose**
  - Ensure that the project product does not contain defects.
  - Get the approvals of authorities.
  - Ensure that operating and maintenance instructions are submitted to the client.
  - Provide information about the main contractor's SCs and other contractors' responsibilities and obligations during the warranty period.
  - Ensure that any training will be conducted.
  - Document experiences.

Coordination meetings for tests and inspections are held to plan the inspection on completion of the work. Who will attend? What are the boundaries for different disciplines—building, HVAC, plumbing, control devices, control electricity, power, IT, etc.? Reconcile planned inspection times and tests.

Final inspection meetings on completion of the work (see Section 6.13.1) are held when the contractor has reported that the work is ready for final inspection. Final inspection begins with an initial meeting where representatives of all contractors, the PM, the SM, the CM, controllers, a representative from the client's operational group, and inspectors are present. At this meeting, the contractors should provide all documentation from previously performed tests.

Before this meeting, the SM, in consultation with inspectors, contractors, and inspectors, has carried out an inspection schedule for the inspection work. (See previous discussion of coordination meetings.) The final inspection is completed by a *final meeting,* where notice of the approval or failure to pass the test on completion is provided.

FAT and SAT meetings (see Sections 6.12.11 and 6.13.1). FAT is held to ensure that the product has no defects when it leaves the factory. It is much better to correct it in the production plant than at the worksite. SAT is held to ensure progressively that the product is delivered without notes of defects. **SAT 1** is a check that all parts are present and of the right quality. **SAT 2** is to verify that the component is working satisfactorily. The **SAT 3** is performed to verify that different systems are working properly with other systems, such as control and regulation of HVAC.

Client's taking over. When the PM is turning over the project to the owner or users, an information meeting should be held to clarify

- Insurance, existing and needed
- Bonds, warranty, and guarantees
- Operating and maintenance instructions
- The project's money situation
- Any validation and classifying regulatory approval
- Possible changes during the warranty period
- Fire safety education
- OSHA issues

Special authority inspections: FDA, radiation safety, agriculture, the Army Corps of Engineers (US), and other agencies. These inspections are performed for some facilities to get them qualified for their business. *Approval meetings* are conducted to obtain permission to deploy and operate services for aviation, ports, railways, subways, tram lines, bridges, and road construction.

Follow-up meetings with users and operating staff are held to

- Inform attendees about contractor responsibilities and obligations during the warranty period
- Compile errors and weaknesses identified but not recorded in the final inspection

Experience meetings with designers and contractors are held to

- Provide information about the experience of both the bid documents and building documents and the contractors' interpretation of them
- Note shortcomings and misconceptions that users believe have occurred during the project
- Discuss good and less than good solutions with designers and contractors

A final reconciliation meeting is held when all financial matters are completed. The PM reports to the client and steering committee.

Guarantee meetings with users and operating staff are held before the warranty period has ended in order to compile the errors and deficiencies found during the warranty period. The comments from users and the management team are documented and included in the warranty inspection. In conjunction with this meeting, the contractors and inspectors are contacted for the guarantee inspection meeting.

Guarantee inspection meetings are held to ensure that the contractor's responsibility during the warranty period ends and any remaining complaints are resolved (see Section 6.13.2).

6.2.3 Coordination

6.2.3.1 Coordination during Design

The DM leads coordination efforts between the consultants. The consultants are responsible for coordination of their own technological solutions with other consultants. However, someone has to control the coordination work if coordinated drawings are to be delivered as scheduled. Aids in the coordination process are borderline interface lists for consultants, contractors, suppliers, and connection managers. Coordination work is done mainly at the DC meetings.

A common coordination problem is collisions between HVAC and electrical installations, building components, and fixtures. Special coordination meetings with the various specialist consultants' documents should always be held, especially for installations in tight places. Each consultant should designate a person in his or her organization to be responsible for coordinating and reviewing documents so that they are complete before they are delivered to the client or contractor. The appointed person's name should be stated in the consultant's project-specific quality plan.

The team of consultants should appoint a person to be responsible for coordination of OSHA issues.

6.2.3.2 CAD Coordination

CAD coordination should be done in accordance with the CAD manual.

6.2.3.3 Coordination during Production

Installation coordination. A project with qualified and complex installations should have a special installation coordinator (IC) with a background as an IM to support the main contractor in the work at the site. Unfortunately, it has often happened that the contractor has appointed a person with no installation experience as IC, resulting in significant production disruptions. The client should prepare a description of the IC requirements, his or her duties, and working time at the construction site. This can be included in the bid request documents and discussed at the negotiations. Coordination of building and installation work is done in both SC meetings and special meetings.

6.2.4 Integration

The project must be integrated with the client's ongoing activities.

6.2.4.1 Designers, Contractors, and Suppliers

Who in the company should work on the project? How will the project be given priority when resources are scarce? Different persons must be assigned responsibility and authority. Sometimes, for formal reasons, it is important that this be in writing. If trade secrets are involved, this may require a personal confidential agreement. How can purchase, transport, and landfill be coordinated with other projects?

6.2.4.2 *The Customer and Users*

When building takes place close to existing operations, it must be determined whether vibrations, noise, electromagnetic activity, dust, machine setups, and even the construction workers' parking could interfere with ongoing operations and accessibility.

When can access to land and premises be obtained? How can existing water and wastewater systems be used? Many times it is appropriate to use existing systems for electrical power, lighting, and heating during part of the building period. Is there capacity for this? Sometimes various media must be turned off in order to connect the project to existing systems. How can this be done? Who should be contacted and how long before the shutdown must the customer be informed? Water, sewage, and heating resources are of great economic value and compensation terms must be clarified in the contract.

What happens in the final stage of the contract or project must be determined. Will the client or users start installing furnishings and equipment prior to taking over? How will the operations and maintenance departments be integrated into the project's final stages? This must be stated clearly in the contract. Will their staff, to a limited extent, participate in the installation, inspections, and test runs? It is appropriate to let operations and maintenance personnel participate in the SAT 3 tests; they should review the as-built drawings and operation and maintenance instructions and comment on them before this testing. If training is included in the contract, how and when should this be done?

6.3 SCOPE, CHARACTERISTICS, AND INTERFACES

WHAT are the functional and technical requirements of the project?

WHAT in the interfaces is included in this project?

WHAT can "fall through the cracks" in a project that has many contract interfaces?

HOW and **WHEN** can information about requirements be obtained and commitments in the interfaces be sorted out?

HOW and **WHEN** will conduct of the coordination be determined and documented so that nothing will fall between the cracks?

HOW and WHEN will interface coordination be stipulated?

6.3.1 General

To establish functional program specifications, an SBS (scope breakdown structure; see Figure 6.4) can be used. (Compare this with the WBS in Section 3.4.1.) To begin, ask **what** the product should do (functional requirements). By asking provocative questions and examining technical solutions, decisions can be made, in conjunction with the users, on the functions and technical standards.

The question is then: **How** can technical solutions to the identified functional requirements be found? There may be special contract requirements and strategies to which it is important to pay attention. There may also be special time requirements. The information is compiled in a scope package, as shown in Figure 6.5.

Information exchange between suppliers of machinery, fittings, etc., and different design consultants can be controlled by using an information exchange template (see Appendix F). (See Section 6.10.3 for more information.)

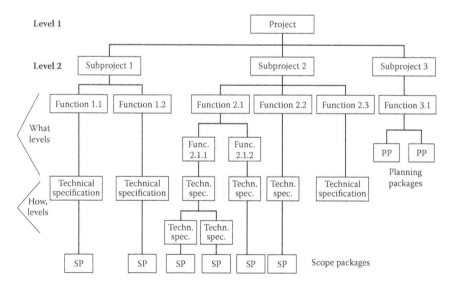

FIGURE 6.4

Scope breakdown structure. Compare to WBS.

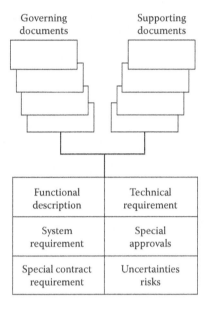

Functional description	Technical requirement
System requirement	Special approvals
Special contract requirement	Uncertainties risks

FIGURE 6.5
Scope package. Compare work package.

6.3.2 Product Description and Design Guidelines

6.3.2.1 Product Description

The product description is a summary of the project's requirements. The requirements are formulated in functional programs, building programs, layouts, key ratio for energy efficiency, and specification standards. These are established in cooperation with the client, users, and designers. An example of key data would be the lighting effect per square meter or air changes per hour. Frequently asked questions in the development of functional programs for rooms are

Ceiling height	Communications	Room function
Floor loads	Special surface course	Floor pits
Daylight	Artificial lighting	Electrical power needs
Climate, temperature, humidity	Cold and warm water	Different types of gas
Risk of spreading infections	Hygienic conditions	Security
Future expansions	Layout flexibility	Telephones, computers
Machinery, equipment	Explosive goods	Sensitivity for vibrations
Electromagnetic sensitivity	Lifting devices	Air quality, allergy risks

As a basis for design work, the client may also have the following governing documents:

- Design guidelines
- Standards for electrical control systems, appliances, markings
- Site master plan for later expansions of buildings and media
- Interface and limits regarding real estate, fixtures, fittings, and equipment

6.3.2.2 Contract Division

- **Purpose**
  - Ensure that the project components (including different contracts) are described or limited and named in a uniform manner. This avoids procurement confusion and results in a simple, uniform filing system.

The client should always choose the best type of delivery or contract and reimbursement. This means taking advantage of everything from functional/performance contracts to divided contracts (see Section 2.5.2) with fixed prices or cost plus. The client (who, of course, is paying) should always be in control of the procurement process, the product requirements, time, and costs. If coordinated general contractor (CGC) contracts are chosen, it is important to describe what the different contracts contain, their limits, and how they will be coordinated. Remember that the client should choose the type of contract that is best regarding cost, time, risk, and quality performance.

Main contractors must also divide their contract into subcontracting and supplies. These must also have limits and clear coordination rules. The deliveries and limits should be documented in a procurement memo. Compare the preceding scope package and procurement packages (Section 6.12.2).

- **Responsibility**
  - Appoint someone who is responsible for writing contract memos when necessary.

6.3.2.3 Contract Boundary Descriptions

- **Purpose**
  - Ensure that nothing "falls through the cracks" when buying various contracts and deliveries in a project.

Since the request documents are not only technical documents, but also legal documents that describe the extent of contract work, it is important that responsibilities be clarified. Early on, who is drawing, describing, delivering, unloading, transporting, storing, installing, and connecting different products must be clarified. In the contract and construction documents, it must be crystal clear who does what. This must be agreed during the design phase and noted in a distinction list. See Appendices G and H.

- **Responsibility**
  - The consultant who describes the work and the purchaser of equipment and machinery should be responsible for informing other consultants of the needs of transport, installation, wiring, etc. That consultant is also responsible for ensuring that an interface responsible list (IRL) is attached to minutes from the PMD meetings.
  - The DM monitors that the boundary is updated.

6.3.2.4 Treatment of the Products Provided

- **Purpose**
  - Ensure that the project's contractor work is coordinated with the delivery of machinery, equipment, etc., purchased outside the project.

Before a facility can be used, special machinery, equipment, furniture, telephones, computer networks, etc., must be in place. The responsibility for this often lies outside the client's building department's responsibility, and the PM is often forced to coordinate this work too.

In addition, many times there is no information about the responsibility for unloading, transporting, storing, installing, and insuring. Who will be responsible for the mandatory receiving inspection? If the contractors on the site should provide any of these services, it must be included in the contractor's contract. This information about contract limits and interfaces should be documented in a distinction list (see the preceding section).

- **Responsibility**
  - The purchasing manager for machinery, furniture, fittings, and equipment is responsible for providing information to the DM in a timely manner.

- The DM is responsible for ensuring that this information is included in the IRL.
- The SM is responsible for coordination on the site if this has not been transferred to the main contractor.

6.4 CHANGE MANAGEMENT

WHAT are needs and what are wants in a request for change?
HOW, by **WHOM,** and **WHEN** can a decision be taken?
HOW will the change affect functional and technical requirements, time, cost, risks?
HOW will everyone be informed about the decided change?
WHEN will the change be implemented?

6.4.1 General

Changes during the project period are inevitable. Changes are also the leading cause of cost overruns and delays. Sometimes the needs change during the project. The project must then be adjusted by making changes. Too often, however, requested changes are based on wants rather than needs. The project manager must have tough control of changes that affect the project's scope, time, and costs. How will time, cost, function, and uncertainties be affected? (See Section 3.3.)

Changes must be traceable and the person who proposed the change and the underlying needs documented. What will the redesign cost? How will it alter the production? What will happen with the schedule? Will it mean changed resources? Who took the decision on the change and is the change within the project's original scope and performance?

Will the change bring more money and time to the project? If a better architectural or technological solution is proposed and this adds costs and delays, the change is not justified if the first solution meets the project requirements. The engineer's or architect's desire to achieve a solution that cannot be done better by someone else must be balanced against the time and cost changes. If the solution meets the project requirements, a redraw will not be done only to satisfy the designer's ego.

During the detailed design phase, between system/basic design and building documents, suggestions of changes occur frequently. These design changes are often based on wants and can have a devastating effect on costs. Changes during the production stage are even more costly. Sometimes we jokingly say that a change in the initial study is the cheapest. The same change during the design phase will cost 10 times as much and during the production phase 100 times as much.

Sometimes the contractor proposes amendments. How will the functional and technical aspects be affected? Who will pay for the redesign or as-built documents? (See Section 6.8.5.)

The PM must determine whether the change depends on wishes or on needs. To keep track of all change requests and decisions, templates can be used, as well as a change log that adopts the change's impact on the cost at completion. Many times, a change proposal to the project restraints (scope, time, and cost) can only be approved by a client or a change control board (CCB).

If a consultant or a contractor has a change to a contract, these changes must be documented in writing. If the client does not write change orders, the consultants and contractors must write a "confirmation of change order."

6.4.2 Change Control

- **Purpose**
  - Ensure that changes are documented and controlled on cost, resource, risk, and time aspects.

6.4.2.1 *Amendment of Governing Documents*

Amendments to governing documents should always be documented in writing and confirmed by the project owner, steering group, client, customer, CCB, or person responsible for the current document.

6.4.2.2 *Amendment of Frozen Documents*

A frozen document during the stepwise design stages forms the basis for the future work of the entire project team. Changing such a document often means additional costs and loss of time for other consultants. Such amendments should be made in PMD meetings. Amendment of bid and building documents must be carefully considered and decided by the PM. Maybe it is cheaper to build according to the documents and implement this change after the performance certificate. Many contractors claim that, because of production disruptions, productivity is impaired. A change log with noted approvals or rejections should be attached to the minutes from PMD meetings. This helps the PM to monitor cost development.

6.4.2.3 Amendment of the Consultant Contract

Changes to the consultant's contract should always be in writing. The PM must write a change order to the consultant on decided changes. If this is not done, the consultant must write a "confirmation of change order" so that, later on, discussions of cost and time changes to the task can be avoided. The question that almost always pops up in the design team is "Who has ordered the design change?—the architect, a user, the CM, the SM, the contractor, or an inspector?"

- **Responsibility**
  - The PM is responsible for documentation of changes to assignments.

6.4.2.4 Modification of Construction Contracts

Changes to contract conditions due to alteration or additional work must be documented in writing. Prepare a template and use it. Cost agreements must be dated and numbered sequentially and separated for building, HVAC, piping, electricity, etc.—for example, B1, B2, B3,…, P1, P2, P3,…, E1, E2, E3,…, and so on. Agreements on additional or reduced costs and time should be sought as soon as possible and signed by both parties. To facilitate payment for additional costs, a copy of the change agreement should be attached to the invoice.

Before a change of work is ordered, the SM should, in consultation with the PM, CM, and TL, control whether functional requirements, the price, and the time are acceptable. Before the work begins, it must also be agreed whether the remuneration should be

- Governed by fixed price
- Governed by unit prices
- Regulated as cost plus, possibly with a price ceiling

Furthermore, note whether the agreed compensation will be indexed or not.

A change control agreement template is shown in Appendix I. The cost control template contains two parts. The upper part of the template contains an order to the contractor with details of how costs should be regulated. The lower part contains a cost breakdown in accordance with agreed payment method. For major changes, it may be appropriate to set the price before work begins. Unfortunately, there is usually no time for this.

- **Responsibility**
  - The SM is responsible, after consultation with the PM, for ordering and documenting change work and all conditions agreed to.

6.5 ENVIRONMENT AND HEALTH MANAGEMENT

WHAT is not acceptable and not environmentally desirable?

WHAT are the environmental standards and targets?

HOW are the environmental requirements and environmental objectives determined?

HOW can it be determined which laws and permits are valid?

HOW are the environmental issues addressed in the project?

WHO is responsible for the environmental issues in the project?

WHAT is not acceptable and not desirable in the working environment?

HOW are the work-environment issues addressed in the project?

HOW is the responsibility for the working environment delegated?

HOW is fire protection for the project controlled during design and production and for the final product?

WHO is responsible for the preceding questions in the project?

WHO is responsible during the design?

WHO is responsible during production?

WHO is responsible for coordinating the work on-site?

6.5.1 General

In this knowledge area, I have chosen to include the environment, working environment (health and safety), and fire protection. The area is often abbreviated SHE (safety, health, environment).

Environmental work is governed by legal codes and company codes. All workplaces must comply with the legal codes for environment and working environment and fire codes. Most companies have a SHE department that specifies requirements and provides advice and guidance in SHE issues. Consideration must always be given to legal codes and the company's SHE regulations, as well as OSHA regulations.

Most companies have an environmental policy and environmental objectives that the project, project participants, consultants, contractors, and suppliers must follow. The PM is usually responsible for ensuring that the laws and regulations are followed and that required permits are obtained before the start of production. Requirements and permits should be incorporated into a project-specific environmental and health and safety (SHE) plan. The PM has to ensure that the objectives and requirements of the environmental and safety plans are passed on to consultants and contractors. This is predominantly done in the form of descriptions and drawings (specifications and building documents). Other environmental studies, aspects, and goals can be added via memos.

In each stage of the planning and design phase, whoever is responsible for a building project (in most countries, the client) should ensure special attention to the working environment during the production phase with regard to

- The location and design of the object or facility
- Choice of construction products
- Choice of designs for the foundation, frame systems, or other structural elements
- Choice of facilities, their location, and attachment
- Choice of fittings, machinery, equipment, etc.

In the production phase, during each stage of planning and design, the owner of the building must ensure special attention to the working environment in the following respects:

- The production period and part-production periods should be generous enough that the work can be performed at such a rate that the risk of illness and accidents is minimized.
- Transport of construction materials, debris, and equipment should be performed in a way that is acceptable from an environmental standpoint.

- The site should be sufficient for the office, staff facilities, etc., without being too crowded.
- Consideration should be given to external vehicular traffic passing by or traffic through the site or area where building work is performed. Vehicular traffic also includes rail transport.
- Prior to demolition, reconstruction, or renovation of an object or part of an object, the risks of hazardous work, material transport, and the stability of an object or part of an object must be studied.
- Depending on the location of the project, the client should appoint a building safety coordinator who has the training, skills, and experience needed to perform the tasks required of a construction safety coordinator under the codes (see OSHA).
- Special attention should be taken when working close to high-voltage cables in air or earth.

The safety coordinator on site should provide information about the risks that may arise from the building activities.

Reconciliation of environmental and safety work should be done regularly—for example, as an agenda item at regular meetings or at separate meetings. The PM often delegates to the SM to decide when and how these reconciliations should be done.

Fire protection is a part of the work environment for the production phase and for the product. This is described in Section 6.5.9.

6.5.2 Environmental Concepts and Definitions

The structure of environment and work environment management is like a pyramid with the policy at the top:

Policy
(Work) environment program
(Work) environment system
(Work) environment assurance plan
Project-specific (work) environment assurance plan
Project-specific (work) environment control plan
Checklists, audits, actions, etc.

Environment management involves long-term and short-term events:

Impact	Environmental	Health and Safety
Long-term effect	Continuous pollutant emissions Depletion of natural resources	Continuous environmental exposure
Short-term events	Environmental accidental emissions Spillage	Fire or explosion Collapse

Definition of Words in This Knowledge Area

Environmental aspect	Parts of an organization's activities, operations, products, or services that may have an impact on the environment
Environmental assurance plan	Plan or program to achieve the environmental objectives
Environmental management system	The part of the overall management system that includes organizational structure, planning activities, responsibilities, practices, procedures, processes, and resources for developing, implementing, achieving, reviewing, and maintaining environmental policy
Environmental policy	Statement by an organization of its intentions and principles for its overall environmental performance that provides the basis for action and defines the objectives and targets
Environmental impact	Any change in the environment, whether adverse or beneficial, that is wholly or partly a result of the organization's activities, operations, products, or services

6.5.3 Common Environmental Objectives

The environmental aspects can be grouped under the following headings:

- Location
- Design work (architectural expression)
- Materials, products, and technical systems
- Functions: air quality, noise, and energy use (can also be included in the preceding headings)

Common environmental objectives for building projects mean that the following statements must be met for a sustainable environmental assurance:

- The lowest impact on health and environment throughout the product's life cycle (in terms of emissions to air, land, and water; energy; and waste) should be used when choosing built-in materials and technical solutions.

- Restrict the use of health and environmentally hazardous products and substances in the building when considering production, operation, maintenance, and repairs.
- Use building materials from renewable raw materials or reused or recycled products.
- Select long-lived building materials and structures.
- Minimize the occurrence of residues during the object's life cycle.
- Reuse and recycle construction and demolition waste in quantities as large as possible.
- Take care of the remaining waste in an environmentally sound way (according to current knowledge).
- Create a good indoor environment.
- Create a good work environment during the production phase.

6.5.4 Legal and Other Environmental Requirements

There are a number of environmental laws, codes, and regulations that often overlap each other. The environmental laws deal with

- Air
- Water
- Land
- Endangered species
- Hazardous waste
- Radiation protection
- Other

A number of regulatory requirements for the construction industry overlap as well.

Remember: Codes are federal, state, and local business. Always check what is valid in the particular area of work!

6.5.4.1 Federal Law

The environment has to be protected against both public and private actions that fail to take account of costs or harms inflicted on the ecosystem. The Environmental Protection Agency (EPA) in the United States is supposed to monitor and analyze the environment and conduct research to devise pollution control policies. The basic purpose of the National

Environmental Policy Act (NEPA) is to force government agencies to consider the effects of their decisions on the environment.

6.5.4.2 Major Federal Laws

A number of laws prevent extinction of endangered plants and animals and seek to recover these populations by preventing threats to their survival (Endangered Species Act). Other laws prevent pollution during the life cycle of the product. This is a manifest system to ensure that waste is properly disposed of and thus not dumped into the environment (Resource Conservation and Recovery Act). The Clean Air Act is designed to protect air quality by regulating stationary and mobile sources of pollution. The Clean Water Act protects water by preventing discharge of pollutants into navigable waters from point sources.

6.5.4.3 Energy Laws

The purpose of federal energy laws and regulations is to provide affordable energy by sustaining competitive markets, while protecting the economic, environmental, and security interests of the United States. US Code Title 42, "The Public Health and Welfare," has many chapters devoted to energy issues and deals with various energy matters. The nuclear power industry is regulated by the US Nuclear Regulatory Commission, whose mission it is to protect the public health and safety from nuclear radiation and waste.

6.5.4.4 Land Use Law

Today, federal, state, and local governments regulate growth and development through statutory law. Natural resources, as defined in the Code of Federal Regulations, deals with land, fish, wildlife, biota, air, water, groundwater, drinking water supplies, and other such resources. If working in or near wetlands is anticipated, it is necessary to obtain a **wetlands permit** (see Section 6.1.1). Check federal, state, and local laws concerning environmental issues that are valid in the area.

6.5.4.5 Working Environment Regulation

The main working environmental organization in the United States is OSHA, which has created a number of regulations. The US Chemical

Safety and Hazard Investigation Board, also known as the Chemical Safety Board or CSB, is an independent US federal agency charged with investigating industrial chemical accidents.

6.5.5 Environmental Activities in the Project

- **Purpose**
  - Ensure that the environment is not damaged by short-term events or long-term effects.
  - Identify significant environmental aspects dependent on project activities and then plan and act in a way so that the company's environmental policy is met.

6.5.5.1 *Process Flow for Environmental Work in a Project*

- **Purpose**
  - Ensure that environmental work is done in a controlled manner (see Figure 6.6).
- **Start-up**
  - Study the client's environmental requirements.
  - Study the environmental requirements in laws; federal, state, and local regulations; and permits.
  - Appoint an environmental manager for the project.
- **Implementation**
  - Conduct an environmental review and environmental analysis for the project.
  - Establish a project-specific environmental assurance plan.
  - Stipulate that the project's environmental assurance plan is a basic design criteria.
  - Establish an environmental audit plan and an environmental control plan.
  - Make an environmental checklist for the project.
  - Make an environmental checklist for the design
  - Make an environmental checklist for production and control that is later incorporated into the contractor's checklist.
  - Consider environmental factors in the procurement of contractors.
  - Analyze environmental risks in the production phase.
  - Review the contractor's environment-related activities in the environmental assurance plan.

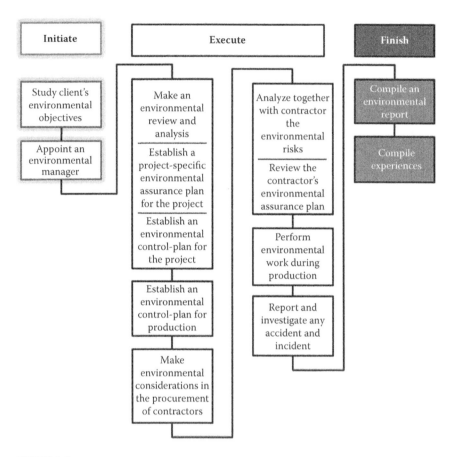

FIGURE 6.6
The process for environmental work in a project.

- • Review the contractor's preparation in case of an environmental accident.
- • Report and investigate any environmental accident or incident.
- **Finish**
  - • Compile an environmental report.
  - • Compile experiences.

6.5.6 Environmental Risk, Health and Safety Risk Management for Future Operations

- **Purpose**
  - • Identify hazards and analyze probabilities and consequences.

A risk analysis should be performed to identify the hazards that exist in the project's future activities. The consequences of an incident and its impact should be investigated. The probability of the event's occurring should also be assessed. The risk analysis provides the basis to prioritize risks and plan activities that makes the risk acceptable. If the consequence is high, an emergency plan should be developed.

6.5.6.1 The Feasibility Phase

Investigate environmental issues in terms of

- Planned building projects
- Activities in the surroundings
- Proximity to water sources
- Past activities on the site because there may be contaminated soil
- Environmental risk in the event of demolition

6.5.7 Establish an Environmental Assurance Plan and Environmental Control Plans

- **Purpose**
  - Ensure that the client's, consultants', contractors', and government's environmental requirements are met.

6.5.7.1 Environmental Assurance

During the feasibility phase, the client establishes a comprehensive environmental assurance plan for the entire project. Environmental assurance plans should then be prepared for the design and the production phases. They should include proposed measures and controls to secure environmentally acceptable products and production methods:

Design phase. Take into account the following environmental considerations:
Design
Material selection
Process methods
Choice of fuel and heating methods
Maintenance practices

Future demolition

Production phase. Take into account the following environmental considerations:

Production methods

Transportation method

Choice of machines

Choice of fuel and heating methods

Emissions to air and land and in water and wastewater

Disposal and storage of hazardous materials

Reusing and recycling waste and construction parts removed during demolition

Reusing building materials and structures under construction

Noise and vibration

Operating phase. Have operating procedures for environmental considerations—for example,

Heating

Cooling

High voltage (electricity and magnetism)

Emissions to air, land (including drainage), and water

Waste

Future demolition. Consider the environmental aspects of

Choice of demolition method

Reuse of material

Hazardous materials

Special environmental areas to consider are

- Impact on natural and cultural environment
- Asbestos
- Contaminated land
- Radon
- Radioactivity (building components exposed to such materials as radioactive isotopes, e.g., hospitals)
- Chemical wastes
- Material and supplies
- Chemical use
- Emissions to air, soil, sewage, and water
- Hazardous waste
- Noise and vibration

6.5.7.2 *Environmental Control Plan and Design Phase*

Project designers should establish an environmental control plan to show the activities and commitments that are required at different stages to guard the general environmental requirements of laws, regulations, and the project's overall environmental assurance. The environmental control plan also sets out the specific targets agreed to in the environmental assurance plan. The guiding principle for the choice of materials is, in particular, to examine the materials present in large volumes and eliminate the presence of substances likely to pose risks to human health or the environment.

Select materials and products

- With a long life
- That are as material efficient as possible
- Made from renewable raw materials
- That can be reused, recycled, or, in the future, used to give energy without harmful emissions (in that order)
- That do not contain environmentally hazardous materials
- Made from recycled materials

A moisture security design check should be performed to determine whether the materials can withstand the moisture load to which they may be exposed. In rooms or businesses with special requirements, any specific environmental requirements should be specified.

6.5.7.3 *Environmental Control Plan and Production Phase*

The contractors set up a project-specific environmental assurance plan and control plan to show the activities and commitment required at different stages to guard the general environmental requirements of laws, regulations, and the project's overall environmental assurance. The environmental assurance plan also defines the specific agreed targets in the environmental policy.

To gain knowledge about the materials and products chosen for use, a list of building products together with environmental declarations for these materials should be created. For chemical substances, a declaration of safety data is added. Material information and safety data should be compiled and sent to the client.

All storage and handling of materials should be protected from additional moisture. All built-in materials should have a moisture content that is below the critical one for adjacent materials. Moisture and dust should not be built into the building. The building must be sufficiently dry and clean to eliminate the risk of moisture damage and microbial growth in the later stages.

6.5.7.4 Construction Waste Management

Many countries have regulations that govern how to handle construction waste. In Sweden, for example, the Work Environment Ordinance/ *Avfallsförordningen* (SFS 2001:1063) must be followed. Generated construction waste should be handled in the following order of priority:

1. Reuse (e.g., windows, doors, machinery)
2. Recycle
3. Energy recovery
4. Landfill

According to Swedish law, the sorting should be done in the following way:

- Hazardous waste
- Electrical and electronic waste
- Combustible waste
- Household waste
- Packaging

Usually, additional fractions, such as metal, concrete and brick, and plastic, are separated at the source.

6.5.8 Work Environment Management in Projects

- **Purpose**
  - Ensure that people are not harmed by short-term events or long-term effects.
  - Establish a process for work environment management in the project.

In the design and during construction, it is necessary to consider how to handle dangerous goods, which include materials that are

- Radioactive
- Flammable
- Explosive
- Corrosive
- Asphyxiating
- Biohazardous
- Toxic
- Pathogenic
- Allergenic

The responsibility for the work environment in construction work projects is often divided between the client and contractors or sub-contractors. The PM must find out what is valid in each case. Has the responsibility been delegated to the PM or anyone else? Remember to consider work environment during production when design is being done. Solutions that mean heavy lifting by hand, working in cramped areas, or injurious working postures should not be suggested (see Figure 6.7).

- **Start-up**
  - Study the client's safety requirements.
  - Study the safety requirements of one's own company.
  - Study OSHA and other regulations regarding work safety.
  - Designate a construction safety coordinator for the working environment during the design phase.
  - Determine any delegation of coordination responsibilities for coordination at workplaces used by many companies (client's production personnel, contractors, and subcontractors).
- **Implementation**
  - Implement SHE analysis for the project.
  - Work the results of the SHE analysis into the design.
  - Establish a project-specific work environment assurance plan (WEAP).
  - Collaborate with the client's fire prevention officer (FPO) in fire protection issues.
  - Obtain opinions from the client's and one's own company's safety officers.
  - Follow the OSHA rules concerning information, permits, etc.

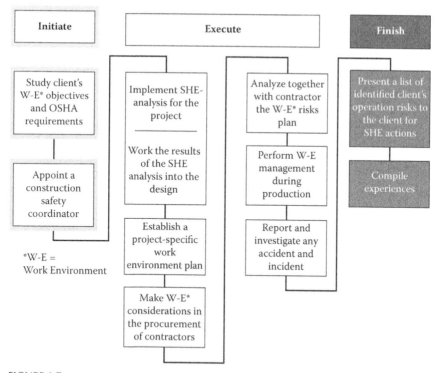

FIGURE 6.7
The process of work environment management in a project.

- Take the work environment into account in the procurement of contractors.
- Take control or delegate to ensure that the responsible contractor's WEAP includes all regulations.
- Perform SHE work, including safety inspections.
- **Finish**
  - Present a list of identified client operation risks and preventive actions to the client.
  - Compile experiences.

6.5.9 Fire Protection

- **Purpose**
  - Prevent employees or property from being hurt or damaged in a fire.
  - Plan fire protection in the finished building.
  - Plan fire protection during construction.

There are a number of fire protection regulatory requirements for the construction industry. Many of these overlap. Information about more regulatory requirements can be found in Section 2.9 in Chapter 2 and Section 6.1.

6.5.9.1 General

Fire safety measures include those that are planned during the design and production of a building or implemented in the structures and those that are taught to occupants of the building. Buildings must be constructed in accordance with the version of the building/fire code that is in effect when an application for a building permit is made. The strategy for fire safety is

- To prevent or reduce the likelihood of a fire
- To prevent fire from spreading to other areas
- To alert those in a structure to the presence of a fire
- To have clear escape routes
- To have a fire prevention officer (FPO) in the company who is responsible for fire safety management

Systematic fire protection. Systematic fire protection work includes organizational fire protection and building physical fire protection. The following elements should be part of good fire prevention:

- Fire protection policy
- Fire protection organization
- Fire safety procedures
- Operating and maintenance procedures
- Design guidelines
- Inspection during construction
- Inspection after contract completion but before moving in
- Fire safety documentation
- Fire safety syllabus
- Feedback of experience

Classification of fires. For appropriate fire protection, it is important to know what type of fire can be faced—such as fires that involve

- Flammable solids such as wood, cloth, rubber, paper, and some types of plastics
- Flammable liquids and liquefiable solids such as petrol/gasoline, oil, paint, and some waxes and plastics, but **not** cooking fats or oils
- Flammable gases, such as natural gas, hydrogen, propane, and butane
- Combustible metals, such as sodium, magnesium, and potassium
- Any of the preceding materials, but with the introduction of an electrical appliance, wiring, or other electrically energized objects in the vicinity of the fire, with a resultant electrical shock risk if a conductive agent is used to control the fire
- Cooking fats and oils because the high temperature of the oils when on fire far exceeds those of other flammable liquids, making normal extinguishing agents ineffective

Structural fire protection includes

- Passive fire protection (fire-resistance-rated walls)
  - Limiting the spread of fire
  - Occupancy separations
  - Keeping fires, high temperatures, and flue gases within the fire compartment of origin
- Active fire protection (manual and automatic detection and suppression of fires)
  - Fire alarm
  - Sprinkler (water or gas)

Design guidelines. The following points must be considered in the design:

- *Fire alarm.* This means both detectors and sounders. The detectors indicate a fire and sounders provide an evacuation signal to the people in the building.
- *Fire documentation.* This describes the conditions and performance of fire protection and the fire protection design. Fire protection documentation shows the building and its components, fire classes, fire cell divisions, exit strategy, HVAC installations, fire protection, and, where applicable, a description of the fire protection installations and a plan for monitoring and maintenance.

- *Flammable goods:*
  - *Flammable liquid* means a liquid that has a flashpoint of not more than 141°F (60.5°C) or any material in a liquid phase that has a flashpoint at or above 100°F (38°C).
  - *Combustible liquid* means any liquid that does not meet the definition of any other hazard class and has a flashpoint above 141°F (60.5°C) and below 200°F (93°C).
- *Explosive goods.* An explosive good is defined as "a material which, when suitably initiated, decomposes with the rapid formation of a large volume of gas at high temperature." Explosives may be solid, liquid, or gaseous and may be single substances or mixtures of different substances
- *Handling of flammable goods.* This includes manufacture, processing, treatment, packaging, storage, transport, use, disposal, destruction, sale, maintenance, transfer, and similar procedures.
- *Hot work (work in a hot environment).* Hot work includes welding, cutting, soldering, drying or heating by flame or hot air, abrasive grinding, and other work that gives rise to an open flame, hot surface, or sparks.
- *Emergency lighting.* This lighting will work when normal power supply fails.

Systematic fire protection work includes

- *Evacuation alarm.* An evacuation alarm is an alarm for hazards other than fires, such as health hazards due to dangerous goods.
- *Evacuation.* A fire escape route is a technically separated path of escape in which no combustible materials may be stored. It should be designed to ensure that any person confronted by fire anywhere in the building, as far as possible, should be able to turn away from it and escape to a place of reasonable safety.
- *Evacuation signs.* Evacuation signs indicate the location of escape routes. The signs can be backlit or illuminated (fitted with electric lighting). In some cases, the signs may also be luminescent (i.e., non-electric signs that are uploaded by the general illumination).

6.5.9.2 The Design Process for Fire Protection

The design must focus not only on fire protection in the finished building but also on fire protection during construction. This is particularly important when connecting to an existing building. Escape routes cannot be blocked and evacuation stairs or temporary lock doors that are escape routes for people in the original building cannot be taken down. A lot of information can be found in the NFPA fire protection handbook.

The process includes the following elements:

- **Start-up**
  - Summarize identified fire hazards.
- **Implementation**
  - Ensure that continuous fire protection measures are included in the design and note them for information.
- **Finish**
  - Transfer the requirements of fire protection equipment not included in the project to the FPO and users.
  - Transfer knowledge of fire prevention measures and fire protection practices that are identified during the planning process to the FPO, users, and operators.

6.5.9.3 Fire Protection in Connection with Occupancy

New construction projects normally include only fixed fire protection installations. Before occupancy, fire protection must be supplemented with movable equipment and training of the personnel who will work in the building. This routine ensures that the building contains or is supplemented with the required fire protection equipment prior to people moving in. This supplement might include the following technical and organizational fire protection:

- Fire protection blankets
- Portable fire extinguishers
- Rescue plans to be distributed to the fire department
- Evacuation plans
- Orientation plans with comprehensive information distributed to the fire department
- Signs for rescue teams' entrance
- Signs that show assembly points

6.5.9.4 Fire Protection Audits

There are two types of fire protection audits: the technical fire safety audit and the organizational fire safety audit. Fire safety audits for the project can be made by the FPO. During production, with limited escape routes and temporary obstacles, it is especially important to have fire safety audits. The technical audit during production is often delegated to the main contractor.

6.6 QUALITY MANAGEMENT

WHAT should be delivered (functional and technical requirements)?
WHAT quality should suppliers deliver?
HOW can the company ensure that it delivers the right quality?
HOW can the company ensure that it gets the right quality from its suppliers?
HOW can revisions and corrections be avoided and work improved?
HOW can it be ensured that temporary work meets requirements?
HOW can the company get the project participants so committed that they themselves control the functional and technical requirements?

A few years ago a study showed that defects in the finished product were due to the following causes:

50%: poor engagement
30%: lack of knowledge
15%: incorrect information
5%: other

The **purpose** of quality management is to

- Ensure that the client gets what is agreed.
- Ensure that consultants, suppliers, and contractors carry out their undertakings in such a way that the scope, schedule, and budget are met.

Management systems for quality control are described in ISO 9000 for product manufacturing and ISO10006 for project management. A new

ISO 21500 for project management in 2011 was a draft version and may later replace ISO 10006.

Preventive actions are always cheaper than correcting errors or rebuilding. A PM plans quality assurance (i.e., identifies quality standards and develops tools in the form of project-specific quality assurance documents, inspection plans, and checklists). Using these documents and processes, the PM will have quality assurance in the project. The PM controls the quality of the project in the design and construction phases by monitoring quality assurance activities and project results and taking action if the results do not meet the quality standards.

6.6.1 Organization

Quality managers in construction projects have multiple roles:

- The client's quality manager
- Project quality manager
- Consultants', suppliers', and contractors' project-specific quality managers
- Building, mechanical, electrical, and special inspectors

A client's QM should control that the quality systems for different departments are effective and applied as intended. The QM of the client, consultants, contractors, and suppliers is a person who is responsible for the company's overall quality management.

A quality manager for the project should control that the quality assurance of the project is effective and applied as intended. The project-specific quality managers of the consultants and the contractors should ensure that the client's quality requirements and their own companies' quality work are carried out correctly in the project. (See Figure 2.2 in Chapter 2.)

Building, plumbing, and electrical inspectors can assist consultants and contractors in the quality work by clarifying ambiguities, monitoring time control, and regulating changes and additional work during production. They can also do spot-checks to ensure that the work is performed as agreed.

The PM is responsible for ensuring that the quality plan is understood and applied by all employees in the project and that the project delivers the right quality. It must be obvious that the discretionary inspections/control of work of the consultants and contractors should ensure that the right quality is delivered, as it is, for example, in the automotive industry, where quality cars are delivered without defects.

Sometimes it may be appropriate to have special quality meetings where the client's QM and the contractor's QM show how the quality of their work is carried out in the project. At these meetings, neither the PQM of the client nor that of the contractor should describe the quality work. The purpose of the meeting is that the contractor's QM—not the PQM—should ensure that the contractor's quality work in the project is maintained. The meeting will conclude with appropriate spot-checks and visits to the workplace.

6.6.2 Different Elements in Quality Work

The structure of quality management is like a pyramid with the policy at the top:

Quality policy
Quality program
Quality system
Quality assurance plan
Project-specific quality assurance
Project-specific quality control plan
Checklists, audits, action, etc.

Quality work in projects contains many elements, including the following:

Receiving inspection. Are the quality and quantity correct?
Discretionary inspections. Is the work done under the right conditions and in the correct way? Was the prerequisite (e.g., concrete moisture ratio before laying carpets) acceptable?
Halfway inspection. Are the project objectives and end-effect objectives met or have they been corrupted during the project work?
Delivery control. Is what has been agreed on delivered to the client.?
Standard/normative inspections. If the client and the contractor cannot agree on what the contract standard is, an independent inspector can "judge." The inspection should be done early in the production to avoid loss of time and money (see Section 6.12.11).

Incident report. The incident report is used to detect whether a prerequisite, product, methodology, or lack of information is or has been a recurring error in the process.

Document management. Are documents archived as has been agreed? How is traceability?

Validation/compliance. Some technical systems and facilities have special requirements. For example, among other requirements, the pharmaceutical industry has a requirement to follow good manufacturing practice (GMP) and good laboratory practice (GLP). Within the nuclear, aerospace, and rail industries, there are similar requirements.

6.6.3 Project Quality Assessment

- **Purpose**
  - Ensure that the agreed quality is delivered to the purchaser (see Figure 6.8).
- **Start-up**
  - Study the client's quality requirements.
  - Study the project's quality requirements, objectives, and end-effect goals.
  - Study one's own company's quality assurance plans.
  - Establish a preliminary project-specific quality assurance plan.
- **Implementation**
  - Establish a project-specific quality assurance plan.
  - Establish a project-specific quality control plan.
  - Prepare checklists.
  - Consultants, contractors, and suppliers adjust their own company's quality assurance plan, control plans, and checklists with the client's documents.
  - Obtain, review, comment on, and approve consultants' and contractors' checklists, quality and inspection plans, etc.
  - Perform discretionary inspections and control of the company's own work.
  - Review and coordinate drawings.
  - Monitor that the quality control plan is met.
  - Carry out spot-tests, audits, etc.

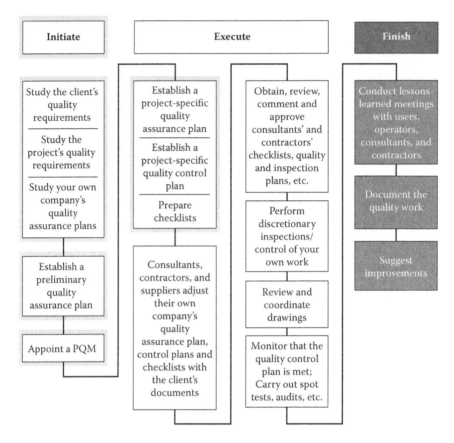

FIGURE 6.8
The process of quality work in a project.

- **Finish**
    - Conduct lessons-learned meetings with users, operators, consultants, and contractors.
    - Document the quality work.
    - Suggest improvements.

Note that the PM or the PQM, in consultation with other project participants, should note if the check is not needed and enter new control points.

6.6.4 Practical Quality Control during Design

- **Purpose**
    - Ensure that the client's quality requirements are incorporated into the design documents.

- Ensure that design documents are complete and coordinated with other consultants' documents.
- Ensure that the required quality and control plans for production are complete before the manufacturing, production, and installation begin.

The consultant checks his or her work

- **When the assignment is obtained** to ensure that the assignment is correctly understood and that the project objectives and end-effect goals are clear and the project's boundaries, times, and payment terms are agreed
- **During the planning**
  - To check and examine the halfway results, which are done through mission briefings with the client at meetings
  - To check calculations with alternative calculations or other methods of calculation or other software, perform manual control (rough estimates) of the rendered data calculations, and compare with previously tested designs
  - To check whether the project objectives and end-effect goals are still being met and have not been corrupted during the design process
  - To review and coordinate drawings with other consultants (It is appropriate to give a special person in the company the responsibility for this work—a person who is not "project blind.")
- **Before delivery** to check and review the final results (This should be performed by personnel with experience and competence. The examination of the documents can be performed by a person who has not worked with the project.)

6.6.5 Practical Quality Control during Production

- **Purpose**
  - Ensure that the agreed quality is delivered to the client.
  - Ensure that the quality assurance work is documented.

The project-specific quality assurance plan of the contractor should include

- Site organization with names, functions, responsibilities, and authorities
- Routines for

- Contacts with authorities
- Communication plan between client, site management, and workers
- Handling of documents, revisions, protocols, etc.
- Receiving and inspection for deliveries to the site, where the number, markings, damage, and necessary documentation on the product's quality are controlled (If responsibility for delivery inspections lies with the supplier, it may be appropriate at this stage to sample and verify that the controls have been done.)
- Control of installation and erection instructions in the beginning of the assembly so that any discrepancies can be corrected before installation has lasted too long; this may imply a revision to or auditing of the contract
- Prototype control
- Ensuring that goods that do not comply are not used
- Coordination of sub- and outside contractors
- Taking action in case of failure or deviation from specified requirements or designs (deviation)
- Project organization and instructions to staff that verify work affecting quality
- Description of how the quality audits are performed

6.6.6 Validation and Compliance

- **Purpose**
  - Ensure that activities with special requirements have quality and validation procedures performed.
  - Ensure that the work is already documented when it begins in the initiation phase.
  - Ensure that production, installation, and documentation are done in such a way that the product can be approved.

The validation process is a method to ensure the right quality and delivery for particularly demanding projects, such as in the pharmaceutical, nuclear, aerospace, and railway industries. During the feasibility study, the client prepares a URS (user requirement specification) that describes the operational requirements for systems and equipment (e.g., ensuring and documenting the continuous pressure of a machine or that the supply air does not contain more than a very small defined number of particles of dirt). Special instructions for production, installation, testing, etc., are described in SOPs (standard operating procedures).

At the same time, a VP (validation plan) is prepared that describes the activities to be implemented from the start of the project until the plant or equipment is used. Before the design can be approved a DQ (design qualification) is usually carried out. This is done to verify that all requirements are met.

During the production tests, inspections and documentation are performed as described in the VP. During the installation, the IQ (installation qualification) is carried out together with FAT and SAT (see Sections 6.12.11 and 6.13). This is followed by functional testing, OQ (operation qualification), and performance under normal production conditions, the so-called PQ/PV (production qualification/production validation). During operation, the facility is secured by CQ (compliance qualification).

All steps in this process must be documented and changes noted to ensure the necessary traceability.

6.6.7 Deviation

- **Purpose**
    - Find product deviations and failures in a company's own work.

Anyone who notes in his or her work that incorrect products or products with deviations are delivered should report this in an incident report submitted to the PQM. Likewise, it should also be reported if the work procedures result in errors that require completion or rework.

It is the responsibility of the quality organization to analyze the incident reports and note if there are failures in the process or if the wrong materials are purchased or used. The analysis may result in

- Action to prevent the emergence of deviation
- Evaluation of preventive actions
- Identifying and taking necessary preventive measures
- Changing processes, procedures, routines, and/or supporting documents
- Education
- More resources

The incident report is used to detect whether a prerequisite, product, methodology, or lack of information is or has been a recurring error in the process. If so, this should be fixed with new or additional procedures.

Deviations can be detected when doing a job, conducting spot-checks, or at quality meetings and during quality audits. Anyone who discovers a discrepancy (the reporter) notes this in an incident report. The reporter notes (if knowledge is available) if the discrepancy is due to his or her own work or, if not, who is responsible for the deviation. The reporter may also propose action to correct the deviation. The report should be submitted to the reporter's company and the PQM of the project.

6.6.8 Quality Audit

- **Purpose**
  - Verify that the quality system is effective and applied as intended.

6.6.8.1 General

There are two types of audits:

- The *system audit* studies the organizational structure, responsibilities, powers, procedures, and resources.
- The *process audit* studies how the various quality processes are used and work.

Audits are usually initiated for one or more of the following reasons:

- To verify that the quality system of a company is in place and that it continuously meets the specified requirements
- Under a contractual relationship, to verify that the quality systems of consultants, contractors, or subcontractors are in place and that they continuously meet the specified requirements
- In order to assess a possible supplier when the wish is to establish a contractual relationship
- When a special event has occurred
- To assess an organization's own quality system against a quality standard

Audits can be done routinely or may be indicated by significant changes in organization, quality, processes, or quality of products or services or the need to follow up on corrective actions. The audit should include questions about the problems encountered so far in the project.

6.6.8.2 The Audit Process

- **Start-up**
  - Define the requirements for each individual audit.
  - Plan the audit and prepare working papers.
  - Establish an audit plan.
  - Send out notices with information about what is to be revised.
- **Implementation**
  - Start up the audit with an initial meeting.
  - Perform the audit.
  - Have a closing conference with the persons who will be responsible for the actions to correct the deviation and discuss and decide when the actions should be completed.
- **Finish**
  - Promptly report critical change to the QM of the audited organization.
  - Report any significant obstacles encountered during the audit.
  - Report the results of the audit without unnecessary delay.

The audit is compiled in an audit report with an attached audit log and a list of deviations. The deviations can be serious, requiring reauditing. If the deviation is easy to correct, a note in the audit log when the discrepancies are corrected (closed) is acceptable. Other observations that can improve the processes should be noted. Who is responsible for the correction and the date when the correction is to be completed should always be specified.

To support the quality work and ensure that nothing is forgotten, many companies have established checklists. These checklists should be adjusted with project-specific points.

The audit should include questions about problems that have arisen in the project. It should be noted whether, according to project members, any of these problems could have been reduced (e.g., if checklists had been used).

The audit process should not be abused.

MEMOS FROM AUDITS SHOULD BE CONFIDENTIAL TO PERSONS OUTSIDE THE REVISED PROJECT AND THE CLIENT.

Neither oral nor written comments should be disseminated without the approval of the revised project team and the client.

6.7 TIME MANAGEMENT

- **WHAT** is the time limit for the project?
- **WHAT** products have long delivery and turnaround times?
- **WHAT** are the critical tasks to get the project finished on time?
- **HOW** can it be ensured that work is progressing in such a way that the project will finish on time?
- **HOW** can planned activities (and which ones) be adjusted to meet the agreed date of completion?
- **HOW** can the uncertainties in time estimates be addressed?
- **WHEN** are decisions needed, request documents sent out, and work or products ordered?
- **WHEN** must certain activities be finished in order to coordinate with other project interests and interfaces?
- **WHEN** must production or installation not take place due to bad weather conditions?

6.7.1 General

The schedule is an important part of planning and involves most of the knowledge areas. The time schedule is one of the main instruments to control a project. The schedule is an activity plan that shows when to start doing a task, the task's duration, and when the activity must be completed.

The purpose of time management is to

- Ensure that the client gets what is agreed to on time
- Select methods of production within the planned production times
- Know when products with different delivery times must be ordered so that they will be delivered in time
- Through resource management, less and over time, reduce the cost of temporary peak loads
- Know if and how to plan activities (on the critical path) if the schedule is not followed or know how much time activity outside the critical path can be allowed to slip without changing the day of completion

- Know when to send documents to authorities, boards, clients, etc., to obtain approvals and decisions in time so as not to cause a delay of completion

During the design and production phases, meetings to monitor how the work is proceeding are held. Is it possible to "plod" on or must the date of completion be honored?

Today, work is primarily controlled with four different schedule types:

- The **milestone plan** shows the big picture and critical dates (milestones) that must be met—for example, approved construction documents, start of excavation, water-tight building, final inspection, start of moving in.
- The **Gantt chart** graphically shows the period within which an activity will be done. This chart is an excellent tool for reconciliations and information.
- The **network diagram** shows the structure/flow chart for the project. Activities are shown as blocks and the logical links and restrictions as arrows. The network diagram is the PM's tool for scheduling optimization, resource allocation, replanning, etc. It provides answers to whether an activity delay will result in the delay of the date of completion or not.
- The **flow-line diagram** for site work shows the periods when many activities are going on in the same areas. The diagram helps to avoid "work jams" in the same workspace.

Other types of schedules are the **critical chain, spreadsheet,** and **project calendar,** but these are not covered in this book.

In the market there is much project planning software, such as MS Project and Primavera. A person is not a planner if he or she can cope with these programs, even if the good-looking plans may induce someone to believe that he or she has project control. It is necessary to know and understand how the programs are working—not to suboptimize. When this is understood, the programs are excellent tools.

Hofsteader's law states:

> *"It always takes longer than you think, even when you take Hofsteader's law into account."* Another law reads: *"A project takes as long as you give it."*

These expressions may be playful, but "there is no smoke without fire." Watch out if there is suddenly more time for a project than is needed. Remember the project triangle. Neither side can be changed without any of the other sides being affected. More time is usually burdened with indirect costs and perhaps lower productivity among employees.

6.7.2 Definitions of Scheduling Terms

Effort: labor units required to complete an activity

Working time: time period (hours, days, weeks…) needed if normal working hours and time for sleep are considered

Milestone: significant event or delivery; has no duration

External milestone: milestone governed by dates not controlled by the project team (board meetings, contract dates, permits)

Internal milestone: milestone decided on by the project team to be able to control the development of the project (start excavation, start erection of superstructure, start installation of special equipment)

Toll gate: an external or internal milestone that requires approval before the work can continue (approved basic design)

Activity: element of work that is required by the project; can be subdivided into tasks; includes time and costs

Relationship/restriction: shows in which order activities can be done; foundations must be done before erection of superstructure

Successive planning: detailed planning for the first period and outline planning for the rest: production cannot be planned before basic design is ready

Compelling logic: an imperative sequence of activities: foundation before erection of superstructure

Preferred logic: sequence of activities based on tradition, but alternatives exist

External relationships: relationships between project relationships and activities outside the project

Lag: a forced amount of time that does not consume any resources: concrete curing before striking formwork

Lead: opposite of lag; an activity can be started before the prior one is complete: painting/laying floor (laying floor on the second floor can start before all painting in the building is finished)

Total float (sometimes called slack or just float): amount of time a start of an activity may be delayed without delaying project end date

Free float: amount of time a start of an activity may be delayed without delaying the subsequent activity

Critical path: the series of activities that determines the earliest completion date; a delay on the critical path means a delay of the completion date; there can be many critical paths in a network diagram

6.7.3 Time Control Process

Detailed schedules are done gradually (see Figure 6.9). As more information about the project becomes available, a more detailed schedule can be made. There is no point in detailed planning work on the building's structure before it has been decided whether to build in situ or use precast concrete. Perhaps the schedule will tell the method to use (see Figure 6.10).

The client's schedules look different from the contractors'. The contractor's production schedules and detailed schedules for complex parts are, of course, much more detailed. Too often, however, a construction schedule shows mechanical and electrician activities as one long activity. How can a schedule like that be monitored?

Before electrical work in a transformer or a telecommunication room can be started, the room should be painted and have lighting and doors with locks. If this has not been clarified in the contract between the builder and the electrician, it can be difficult for the electrical contractor to start work as planned. This will result in uneven requirements of labor and force the electrical contractor to add resources or work overtime, which often leads to additional costs for the electrical contractor. A detailed schedule will show critical relations between different contractors' work. The schedule must also show when large ventilation equipment should be placed in the air handling room.

The subcontractor's schedules must be broken down to a level where the work performance can be monitored.

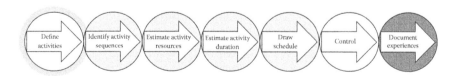

FIGURE 6.9
Time control process.

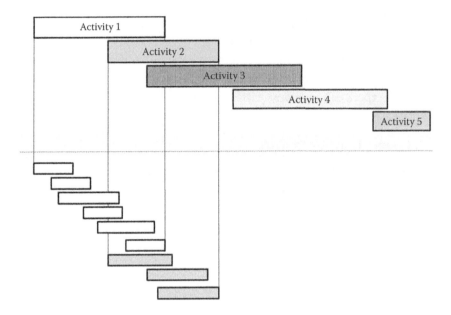

FIGURE 6.10
Successive planning.

6.7.3.1 Milestone Plan

Scheduling normally starts with a comprehensive flowchart—a milestone plan. Important events and deliveries (milestones) should be identified. In all projects, even small ones, some milestones should be identified. The number varies between projects, but it is normally between 4 and 20. Usually, one starts from the end of the project and notes what must be done on the way. This method is called "back-casting." The result might look similar to Figure 6.11.

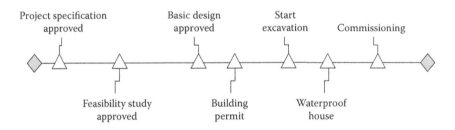

FIGURE 6.11
Example of milestone plan.

Note that no dates are yet selected. When consensus on the milestone plan is reached, maybe after a few additions, then defining the activities needed for scheduling can begin.

6.7.3.2 Defining Activities

Based on the project charter, contract, project plan, and previous experience jointly developed (e.g., through brainstorming), the various activities needed to identify milestones are defined. This can also be done structurally by developing WBS, what, when, and how. Work packages and planning packages are created. If one has worked with WBS, what needs to be finished before a task is started has also been identified. One or more relations or restrictions have been identified.

Often, there are templates that support the task decomposition. Just remember that every project is unique and a template or a checklist is only an aid. It does not contain all activities. Determine the activity sequence and present it in a precedence network diagram. There are basically four alternative task sequences (see Figure 6.12):

- **Finish-to-start.** A task must be completed before the next can start.
- **Finish-to-finish.** One activity must be finished before another can be finished.
- **Start-to-start.** One activity must start before another can start.
- **Start-to-finish.** One activity must start before another can finish.

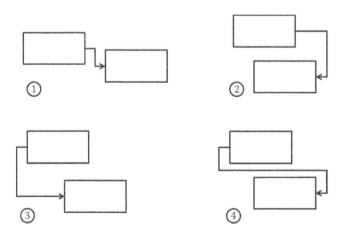

FIGURE 6.12
Alternative task sequencing.

A graphic image in the form of a precedence network diagram is compiled from the defined activities and their restrictions.

6.7.3.3 Determine Resource Requirements and Duration of Tasks

In this process, experience and knowledge of available resources are needed as well as special equipment. Experienced PMs, consultants, planners, and contractors have data that can help. In the market there are also books with data that can be used to support work at this stage.

Often, the duration is expressed as "10 days ± 2 days," or "at least 7 days," or "probably 10 days," or "not longer than 14 days." Of course, the uncertainties are great in the beginning of the project, but with the help of, for example, *PERT* (Program Evaluation and Review Technique; see Section 6.11.8) or the *Lichtenberg method,* the project team can identify *weighted average durations* for various activities. With the generated data, a schedule such as the Critical Path Method (CPM) can be established.

6.7.4 Network Diagramming

If how the computer works with CPM is not understood, it is easy to suboptimize. To understand this, a simple case will be described. Of course, in real life, the forward pass, backward pass, etc., would not be counted because the computer does this. But it is important to understand the principles and what "floats" means and how the critical path moves in the block plan when delays occur.

As a base, a network diagram like the one in Figure 6.13 is used. The network can be analyzed by *PERT* or the *Lichtenberg method* (statistical sums) or simulated by the *Monte Carlo model.*

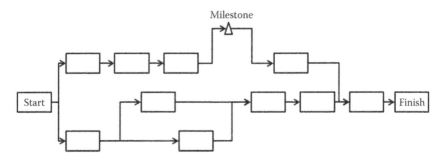

FIGURE 6.13
Precedence network diagram.

6.7.4.1 Forward Pass

A network diagram should always start with a start box and end with a finish box. The task's name (A, B, C,...) and duration (e.g., number of days) are written inside the activity box. The number of days should be the "weighted average durations" according to PERT or the Lichtenberg method:

PERT weighted value = (pessimistic time + 4 × likely time + optimistic time)/6

For more about PERT, see Section 6.11.8. To evaluate the impact of uncertainty on the project, see Section 6.7.6.

The computer program begins with a so-called *forward pass,* the earliest possible start and earliest possible completion for each activity, as shown in Figure 6.14. A4 stands for task A, with an effort of 4 days (*weighted average durations*), B7 is the activity B with a weighted average duration of 7 days, etc. The earliest possible start of the activity is written in the upper left corner and the earliest possible finish in the upper right corner. Activity A can start no earlier than on day 0 and be completed no earlier than day 4. Activity B can start no earlier than day 4 and, at the earliest, be ready on the 11th day. Activity E can start no earlier than day 4 and, at the earliest, be completed after 19 days, etc.

The fastest time that this project can be implemented with selected resources and weighted average duration is thus 42 days. The computer calculation is as follows:

- Begin in the starting box and move from left to right.
- Enter the "earliest start for activity j" (ES$_j$).

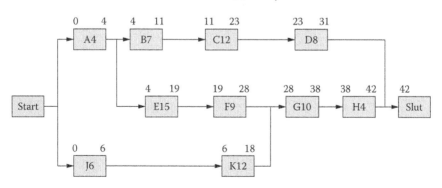

FIGURE 6.14
Forward pass.

- Add the task's duration, which will yield the "earliest activity-finish" of activity j (EF_j).
- ES_j + weighted average duration = EF_j.

For multiple relationships, the ES of the subsequent activity is the latest EF. When lag (delay) and lead (overlap) are taken into account,

$$ES_j = EF_i + lag - lead$$

where ES_j stands for early start activity j and EF_i is the earliest finish of a premature activity.

6.7.4.2 Backward Pass

The computer continues by making a backward pass (see Figure 6.15). One looks for the latest finish and latest start for each activity that is acceptable if one should finish within the number of total days found in the forward pass. The numbers are written below the squares.

In the figure, task H must be completed in no later than 42 days, which means that it must start no later than day 38 (42 – 4). Task D must also be completed by day 42. As the activity will take 17 days, it must start no later than day 25; if not, the end date will be delayed (42 –17), and so on. If 0 appears in the start box, counting has been incorrect.

The computer calculation is as follows:

- Begin in the finish box and move from right to left.
- Enter the "latest finish for activity j" (LF_j).

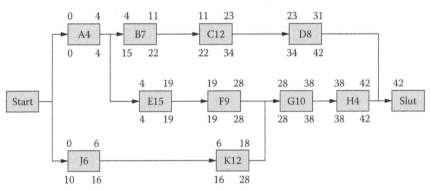

FIGURE 6.15
Backward passes.

- Subtract the task's duration, which will yield the "latest activity-start" of activity j (LS_j).
- $LS_j = LF_j$ – weighted average duration.

For multiple relationships, the LF of the preceding activity is the latest LS of a subsequent activity. Taking lag (delay) and lead (overlapping) into account,

$$LF_i = LS_j - lag + lead$$

where LFi stands for late finish for activity i and LS_j stands for the latest start for activity j.

6.7.4.3 Floats and Critical Path

The next step in the calculations is to calculate the float. This is done by taking the difference between LF and EF (or LS and ES) of a task. In Figure 6.16, activity D has a float of 2 days, activity F a 0-day float, and activity K a float of 10 days. Where the gap is 0, the critical path is present. A delay of an activity on the critical path causes a delay of the final time. The start of activity D can be delayed by 11 days in relation to its early start date without delaying the end date. This is called the total float (or just float or slack).

Looking at the activity of K, it can last for 22 (12 + 10) days without an early start for the subsequent activity of the G10 will be delayed. This is called free float. How many critical paths can there be? What

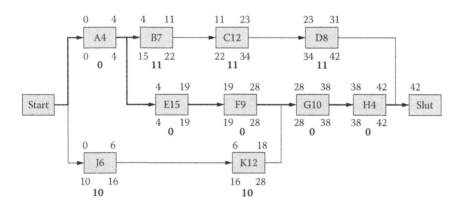

FIGURE 6.16
Floats and critical path. Fat figures are floats. Thick arrows indicate the critical path.

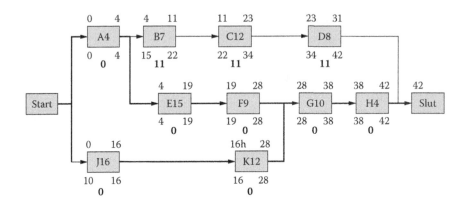

FIGURE 6.17
Network plan with two critical paths.

does it look like if the weighted average duration of K is 22 days (see Figure 6.17)?

There are two critical paths.

Is it complicated? Do not worry. Do not forget that the computers do the calculations. It is just necessary to fill in the weighted average duration of each activity, fill in the lags and leads, and specify how activities are related.

Now that there is a network diagram, with one or more critical paths and the knowledge of floats, analysis can begin. Does one have 42 days at one's disposal? If not, it is necessary to act. There are usually three ways to react:

- *Crashing*: shortening durations, adding staff, working overtime, or using more machines
- *Fast-tracking*: looking for some activities that can be parallel instead of consecutive
- *What if*: examining the opportunities and threats that exist if the work is not finished in time

It is not enough to ensure the total project length. External milestones and long holidays must be taken into account. Transfer the network diagram activities into a Gantt plan containing the calendar dates. How does it look now? Are specific dates for external milestones (board decisions, political decisions, etc.) at stake? The tasks that must be forced or "fast-tracked" can be analyzed. Now, when the final date has been

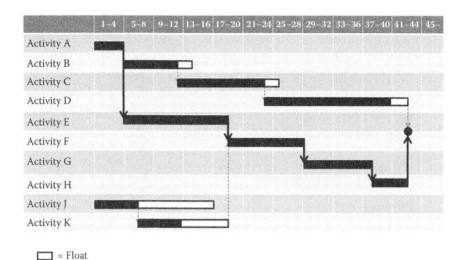

FIGURE 6.18
Gantt plan with critical path.

determined, the milestones must be checked. The critical path may pass a milestone—what does that mean?

Activities with floats are more difficult to transfer. Normally, they are identified by an early start. If there are floats, the duration of the action can be extended or the start delayed. This will be used when resources are optimized.

The information in Figure 6.14 gives the Gantt plan in Figure 6.18. Insert the milestones in the Gantt plan and check how it goes. Does any part need to be speeded up?

6.7.4.4 Crashing and Fast-Tracking

By increasing the effort of an activity on the critical path, the total time can be shortened. Note that it is necessary to "crash" with one unit at a time. Changing with several units can lead to suboptimization as a new critical path takes over. See the following example with the network diagram from Figure 6.15.

If the goal is to be completed in 38 days, it is proposed to reduce the effort of G in the figure by 4 days as it lies on the critical path. Will the project be finished in 38 days? No! The critical path has moved to the activities of A, B, C, and D and the project will take 40 days.

If the activity of E is reduced by four units and of D by two units, will the project be finished in 38 days? No! The critical path of J, K, G, and H is left with 42 days. It is necessary to keep in mind to take one unit at a time and study what happens to the critical path. The computer can do all the forward and backward passes, work out floats, etc. But it is necessary to understand what happens with the critical paths. The same reasoning must apply when working with fast-tracking.

6.7.5 Resource Optimization

- **Purpose**
  - Distribute mechanical resources and manpower resources in an optimal way.

A construction project requires resources in terms of people and machines. Sometimes special machines and people with unique expertise in various projects are needed. When are these special resources needed? Are they available when they are wanted? It may be necessary to negotiate with other projects or a program director. Is it possible to move the time when the resource is needed? Are there floats?

Of course, it is desirable to have a uniform distribution of resources during the project time. "Peaks" means additional costs in learning and costs for renting temporary equipment and personnel. In order to clarify the resource issues, the resources are studied with the help of the network diagram, which is summarized in a histogram (see Figure 6.19).

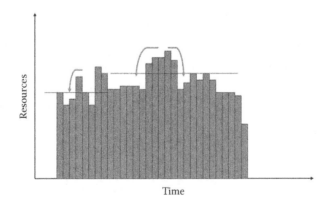

FIGURE 6.19
Histogram of resources.

If there are floats, the information from the histogram can be used to delay the start of various activities. A more balanced histogram may result.

6.7.6 Uncertainty Analysis

- **Purpose**
  - Clarify how uncertainties affect different times.

Effort estimates include uncertainties. The most likely, the fastest, and the longest durations have been obtained. A network diagram and PERT can help to evaluate the critical path. (For more information on PERT, see Section 6.8.11.) If the standard deviation and variance for each activity are calculated, the standard deviation for the critical path of the entire project period can be studied (see Figure 6.20).

With the help of "weighted durations," the "most likely project duration" is found to be equal to $4 + 15 + 9 + 10 + 4 = 42$. Standard deviation for the critical path is $(1.33^2 + 1.33^2 + 0.33^2 + 1.18^2 + 0.33^2)^{0.5} = 5.3$. With 84% certainty, the project is done within 48 days $(42 + 5.3)$ and, with 95% certainty, within 53 days $(42 + 2 * 5.3)$. This is important information for planning and contract signing.

PERT can only be used for the various critical paths and not for the uncertainties of other parallel activities. For that, the Monte Carlo model can be used for simulations. With this model, various activities can also have different statistical distributions.

6.7.7 Monitor, Act, Replan

At regular meetings, the schedule is monitored and where the project stands is evaluated. Is it ahead of or behind schedule? DO NOT estimate

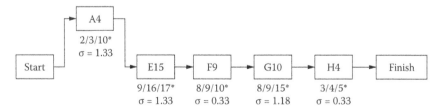

FIGURE 6.20
Uncertainty analysis of the critical path.

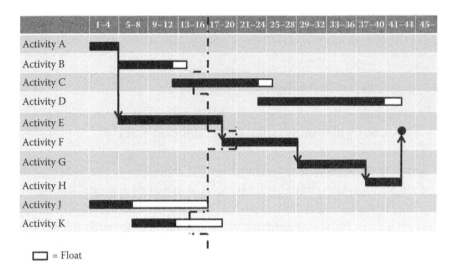

FIGURE 6.21
Monitored Gantt plan.

what has been done! To get the current mode, estimate how long it takes to complete the task and count backward from the end of the action. If there is a delay on the critical path, it is necessary to act. The next time, the revised schedule is monitored against. To do nothing is to conduct the project incorrectly.

Today, a straight line is used—not an irregular line as shown in Figure 6.21—for the monitoring date and to see what happens with the activities when the numbers for "time to complete activity" are input. Performed work is filled in by blackening the activities. It is also necessary to check the original Gantt schedule (Figure 6.22) that meets all milestones to know how to act to meet the agreed date of completion.

In the Gantt plan, two activities need action. Notice also that activity F could start before planned and did so. It was possible to do some parallel work that was not planned when scheduling the first time.

6.7.8 Flow-Line Plan/Site Disposal Schedule

A site schedule or diagram (flow-line plan; Figure 6.23) shows periods during which various activities work in the same area. This helps to avoid labor jams. Areas where many activities are going on at the same time must be studied in more detail. This may result in schedule revisions.

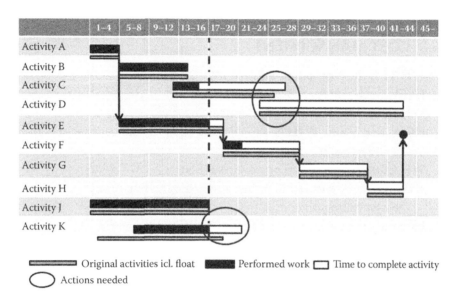

FIGURE 6.22
Monitored time schedule.

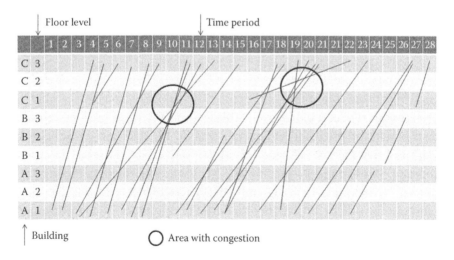

FIGURE 6.23
Flow-line plan with areas of congestion.

6.8 COST MANAGEMENT

WHAT is the cost ceiling for the project?

HOW much will it cost?

WHAT and **WHEN** will payment be?

HOW much and **WHEN** do salaries, supplies, subcontractors, etc., need to be paid?

WHAT is it necessary to borrow? **WHEN?**

HOW can it be ensured that the work is progressing in such a way that the cost ceiling is adhered to?

HOW are the uncertainties addressed in cost estimates?

6.8.1 General

A decent life is based on a controlled money situation. Whether the company survives or dies depends on that situation. Overinvestment or a bid that is too low may result in bankruptcy. Too highly anticipated bids mean losing the contract, and this results in employment cuts. Remember that salaries come from living companies.

Cost management has many subprocesses:

- Estimation
- Building a cost reference plan
- Building a cash-flow plan
- Cost control

Investments and bids are based on calculations. These must be based on good knowledge, but this is not enough. There is great uncertainty when a project is started or a bid is given. This leads to risks that must be valued, distributed, and priced. It is important that the PM take part in the budget work and that the CM, TL, and IM take an active part in the bidding process and negotiations. Project knowledge is the key element to identifying quickly what should be done and what the estimated cost will be. One must also be able to see whether the project is on its way to derailing and act upon that fact.

Knowledge about the project's money situation comes through analysis, progress reports, and projections. Within the cost segment, the following interactive elements are mainly worked with:

- Cost assessment
- Budgeting
- Cost control
  - Progress reports
  - Forecasts
  - Progress analyses
- Reactions and active decision making

To assess profitability, it is necessary to know investment costs. When a bid is submitted, what it costs to perform the work, reserves for the unexpected, and a small or big profit must be evaluated. It is necessary to make cost estimates.

During the project, the PM, TL, CM, and IM must analyze the cost development of their projects or contract. How is it going? Does it look good? As a PM, TL, CM, or IM, do I have to act? To know this, it is necessary to have something against which to compare. With a budget for the project, the reference document for budget costs is created; later, the current situation can be compared against it.

Today, many different tools are used to analyze and forecast the cost situation in a project. Managers need to have earnest and true financial status reporting to feel comfortable and not start to interfere with the project work. The cost management processes are illustrated in Figure 6.24.

6.8.2 Cost Estimation

The first qualifying cost assessment is done in the initiation phase of the feasibility study. The uncertainty of this is large, of course. In some projects, it may be as large as –25% to +50% and in others smaller (–10% to +15%). When we retrospectively studied projects where the actual cost was higher than the original estimate, it can be said that this usually depended on

- Unclear requirements on the scope and performance
- Items that were forgotten when calculating was done
- Needs and wishes that have been out of project control
- Delayed project
- Market forces that are not working properly
- Many alterations and additional work because the planning, design, or procurement has been poorly implemented

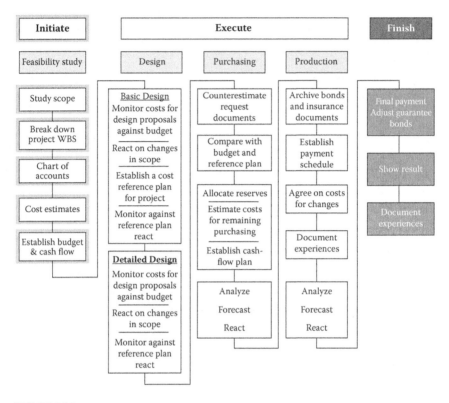

FIGURE 6.24
The process of cost control efforts in a project.

- Uncertainties that were neglected
- Cost index development
- Currency fluctuations

Remember that there are direct and indirect cost components:
Direct costs (directly attributed to the project) include

- Labor
  - Internal
  - Contracted
- Material and equipment
- Other direct costs
  - Travel
  - Fees
- Other

Indirect costs (overhead) include

- Administrative, general
  - Headquarters expenses
  - Benefits
  - Telephones and electricity
  - Heating
  - Fringe benefits
- Marketing and sales
- Research and development

6.8.2.1 *Planning the Calculation Work*

Before the calculation process begins, the following must be determined:

- Level of accuracy
- Whether the description of the project scope is good enough
- Which parts of the calculation can be done by the company and where help will be needed
- Which parts of the production should be done by the company and what should be bought
- Who is responsible for the calculation work
- How index changes and currency fluctuations should be handled
- Type of remuneration
- Payment schedule
- How the project will be financed and whether a cash-flow plan is needed
- Who should receive information about bids, budget, and the monitoring of costs

6.8.2.2 *Calculating*

Input for the calculation process includes the project specification, the company's historical data, checklists, charts of accounts, WBS descriptions, resource plans, volume calculations, market knowledge, and a risk list. The level of knowledge of the work determines the calculation method. A contractor should divide the contract into subtasks. The documents, drawings, and descriptions are a good basis for the calculations. The PM has much less detailed information for the project

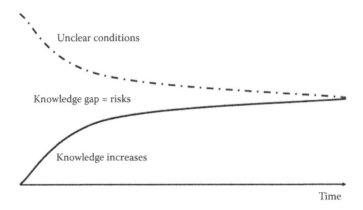

FIGURE 6.25
Project knowledge and uncertainties.

calculations. Calculating work must be more detailed as the project gets more specific.

Both the client and the contractor must take into account the uncertainties (risks) that exist in their mission. There is a *uncertainty gap* that will decrease as the project moves along (see Figure 6.25).

In early stages when there is limited knowledge of the project, *top-down calculation* is used. This means a rough estimate—that is, to compare similar projects (referenced assessment). It is important that the projects be similar and that calculators be knowledgeable about this type of calculation. It is necessary to concentrate on the differences between the earlier project and this one, and how the market has changed.

If there is good knowledge of the project, *bottom-up calculation* is used. Detailed calculation is based on "safe" quantities and prices. Preliminary bids from potential suppliers and subcontractors have been obtained and quantities and hours of work tasks calculated. If the company does not have reference data that show unit cost (running meters, square meters, cubic meters, watts, air change rates)—so-called *parameter calculation*—there are consulting firms and manuals that have information on this.

6.8.2.3 Reserves for Uncertainty

The calculation must take uncertainty into account by creating financial reserves. A calculation method that directly addresses the uncertainties is the *Lichtenberg method*. Similar procedures that take the uncertainty in various parts of the calculation into account are to use a weighted PERT value or the simulating method of the Monte Carlo method.

With the *Lichtenberg method* and PERT, the minimum, maximum, and most probable costs of each part to be calculated are estimated. Mathematical formulas give the weighted value and the standard deviation. The *Lichtenberg method* continues to determine the variance. This is done for all cost centers. The cost center with the largest variance is prioritized and broken down into smaller components. Then each cost position is estimated: minimum, maximum, and most probable values. New estimates of standard deviation and variance yield new priorities and the cost centers with high variances are broken down into smaller parts. This is repeated until an acceptable cost level is reached.

There are two types of reserves: contingency reserves and management reserves. The first takes into account the identified uncertainties and can be measured with different tools. The management reserve is for nonidentified uncertainties and is normally a percentage figure—sometimes a company policy figure.

Some uncertainties have such an impact on the project that it is necessary to take concrete measures to reduce the likelihood that they will occur and to have an emergency plan if they do. Of course, the cost for these risk-reducing actions must be included in the budget and not in the reserves (see Figure 6.26).

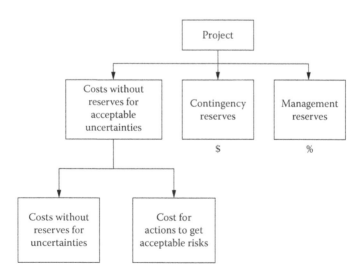

FIGURE 6.26
Knowledge project and uncertainties.

When a dispositive budget plan is used, the contingency reserve should be allocated to the cost center in the chart of accounts, while the management reserve should have its own cost center.

6.8.3 Cost Reference Plan

The time schedule and the budget reference plan or the dispositive budget plan are two main reference plans against which how the project develops is assessed. Many of today's budget reference plans in the construction industry (clients, designers, subcontractors) are based on the way they were calculated. Such a budget cannot be used as a reference plan.

Input to the budget process consists of the project specification, data, chart of accounts, checklists, WBS descriptions, cost estimates, and risk list. With the budget and timetable, a graphical representation of costs over time is established (cost baseline, S-curve).

The sum of the accumulated costs will be added as a reference curve that can be used for monitoring during the implementation phase (see Figure 6.27). It can also help to determine funding needs (cash-flow) for the project if the schedule is followed and the actual costs are the same as what was calculated.

6.8.4 Cash-Flow Plan

Wages and bills are payable and there must be money to pay. If the money is not available, it is necessary to borrow. With the help of the reference plan (see Figure 6.27), one can assess when it is necessary to borrow and how much needs to be borrowed. The difference between costs and allocated funds must be financed. Interest is assessed on borrowed money. These are costs that are charged to the project without adding any value. A skilled PM, CM, and IM make sure always to have a positive cash-flow.

For a contractor, who generally gets paid only after the work is undertaken, it is important to include the interest expense in the budget and keep interest costs as low as possible.

In order to know how money gets into the project or to the contractor, it is necessary to include a payment schedule in the contract.

In Figure 6.28, the "payment for wages and bills" is located over the "payments for executed work" (dashed line), which indicates how much money must be borrowed. In many negotiations, the contractor or supplier tries to get a "front-heavy payment plan" (dashed-dotted line),

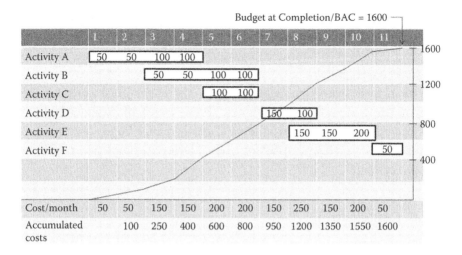

	1	2	3	4	5	6	7	8	9	10	11
Activity A	50	50	100	100							
Activity B			50	50	100	100					
Activity C					100	100					
Activity D							150	100			
Activity E								150	150	200	
Activity F											50
Cost/month	50	50	150	150	200	200	150	250	150	200	50
Accumulated costs		100	250	400	600	800	950	1200	1350	1550	1600

Budget at Completion/BAC = 1600

FIGURE 6.27
Gantt plan with costs for activities.

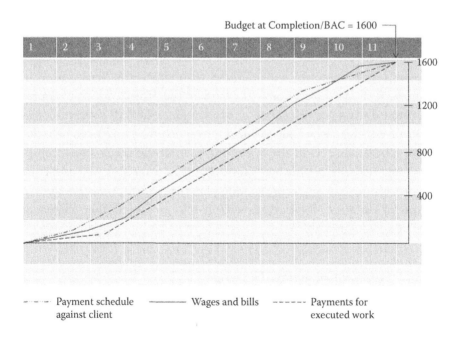

Budget at Completion/BAC = 1600

- - - - Payment schedule against client
———— Wages and bills
- - - - - Payments for executed work

FIGURE 6.28
Cash-flow plan.

which is located over the wage and bills (straight line). If the contractor succeeds, he or she does not have to borrow money and pay interest. Another option is to pay subcontractors a little after payments come to the main contractor.

With a front-heavy payment schedule, the client must consider the risk of the contractor's bankruptcy. These types of payment schedules need constant monitoring, even if performance bonds are in place. Bankruptcies are always costly for clients. Delayed completion results in delayed revenues.

6.8.5 Cost Control

6.8.5.1 General

The cost management process includes cost estimates, economic forecasts, analysis, and action. The money situation is also affected by change management and changes to the schedule. The process includes bond and insurance management and invoice verification. The design consultants should consider their drawings and descriptions as production. The TL must therefore establish a production-driven budget or reference plan (see earlier discussion) for the work.

The possibility of influencing the project is greatest at the beginning. As the project develops, the possibilities become fewer. The work intensity (drawings and building) will increase with time. The black curve in Figure 6.29 shows the economic turnover of the project, which of course is greatest during the last phase of production, when all subcontractors are involved. The dashed curve shows how much the project can be influenced. Of course, it is in the beginning that all the really smart thinking should take place. This holds for the client's project as well as for the contractor's work. A smart production plan in the beginning can save the contractor's money. The dashed vertical line represents the schematic/basic design documents.

6.8.5.2 Cost Control during the Design

During the first stages of design (system/basic design documents), the investment cost versus operating and maintenance costs must be considered. For example, a low investment in insulation and heat recovery obviously means increased operating costs.

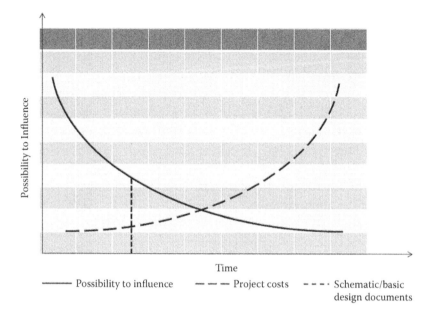

FIGURE 6.29

The ability to influence the project cost (black line) versus costs (dashed line) in a project.

Market forces and competition must also be considered when materials and workmanship are selected. With the system/basic design documents as a base, a refined calculation must be made. Can a technical system be approved without the knowledge that it is within the project's financial framework?

Imagine that, in each PMD meeting, four or more wishes (not needs) of $1,500 each are decided upon. During a period of 4 months, with PMD meetings every 2 weeks, the results of these "small" wishes raise the costs by nearly $50,000. One way to keep track of the cost of these small "improvements" is to set up an expense log of costs and attach this to the minutes from meetings. The study of this log may result in action. For an example of a revision log template, see Appendix J.

6.8.5.3 Cost Control during Procurement

In connection with purchases, the first concrete answers about how the design has been kept within the project's economic framework and how successful the calculations have been can be obtained.

A professional bid analysis requires that the budget for the proposed bid be known. In addition, the request documents should be countercalculated by the buyer before the opening of bids. If the clients analyze the request documents, they will find ambiguities and shortcomings in the documents and get a signal if the market does not operate normally. One can also ask oneself whether one's own cost-calculation tools are good.

The type of remuneration to be used must be decided. Types of remuneration include

- Fixed price with or without indexation and/or incentive
- Quantities and unit prices
- Cost reimbursement
- Different types of incentive contracts

To ensure that the client and contractor can fulfill their commitments, there are a number of bonds, insurances, and guarantees. Are the costs for these included in the budget and bids?

Another economic issue to discuss in negotiations is the payment schedule. What does cash-flow look like? The routines for change management that influence the money situation must be settled during negotiations.

6.8.5.4 Cost Control during the Production Phase (Design and Building)

Changes during the production phase are expensive. If a change will cost $100 to implement during the feasibility phase, it is said that it costs $1,000 during the design phase and $10,000 during the production phase. The cost of the disruption suffered by the contractor in lowered productivity is often much greater than what is later received in compensation from the client.

To change the location of a door in drywall that has not yet been built can be attractive. Can there be any cost for that? To change the drawings, which often involve more than the architect's drawings, can cause a relatively high cost. How does the new location of the door affect the HVAC ducts in the corridor and the different electrical systems? Do the installation drawings need to be changed?

Sometimes the contractor proposes changes; the savings will be shared between the parties. When this happens, ask the question: "Have you included the redesign?" One must control that the change meets the functional and technical requirements, changes drawings and specifications, etc. Many times, the proposed change suggested by the contractor will not be such a great saving.

For project management, it is important to clarify who has the authority to approve changes during the production phase. Be careful regarding "apparent authority." This is an authority that was **not** intended, but by virtue of actions, behavior, or title gives the appearance of genuine authority.

Changes during production will always happen. This does not need to be due to bad management. Soil conditions or other conditions may change over time. The design has been based on a number of assumptions with varying degrees of uncertainty. When changes or additions occur, settle the issue immediately and document what was decided. If the cost cannot be determined immediately, agree instead on how it should be regulated. This benefits both the contractor and client. Try to have regular finance meetings between the client and the contractor.

Throughout the production period, the economic situation should be monitored regularly. Compensation for changes and additional work should be done as soon as possible. Appendix I is a template for economic settlements.

6.8.5.5 Prediction of Cost at Completion

During the project, it is necessary to get an idea of what the cost at completion of the project will be. Will it be within the budget or must action be taken? Have too many "wishes" beyond the project's needs been allowed? If the project cost cannot be reduced, what can be done? Because it is important that the client or project owner have time to act, reporting must take place continuously.

Invoices are no help in predicting cost at completion. It is when the products are ordered that the cost at completion is monitored and documented. This is an ongoing task that should be reported once a week or fortnight. Cost analysis should be done in conjunction with system/basic documentation, before and after purchases, and periodically during production.

Each cost center should be analyzed for

1. Completed purchases
2. Estimated remaining purchases
3. Cost reimbursement/actual costs
4. Cost reimbursement/estimated remaining costs
5. Settled change orders
6. Estimated cost for not settled change orders

7. Estimated reserves for future change orders
8. Cost-at-completion forecast
9. Deviation from budget
10. Budget
11. Invoiced at cost center

Item 11 is included only to ensure that double billing does not occur or that the costs of other projects are not charged to this project.

The number of cost centers is tailored to the project in order to have financial control and get good feedback for future projects. The table may, if desired, be supplemented with columns for "adjustment of the budget" and "new budget."

The PM must work together with the team members and continually evaluate the development of items 4–7 (production phase). The starting point for the assessment of items 4, 6, and 7 should always be "How much is required until completion?" and should certainly not be "This much is left of the contingency reserves."

6.8.5.6 Monitoring Project Performance with Earned Value

The earned value method is a performance/forecasting method that is successfully used in many projects. The method presents performance along time and cost lines in a clear way.

Consider the following: The design has been planned, a schedule made, and a budget created that is broken down month by month. If, after 3 months, more has been consumed than was budgeted, will there be a budget overrun? It depends. Is the project ahead of schedule, just as planned, or behind schedule? Perhaps the project is far ahead of schedule and less has to be spent than was thought at this point. Simply to study how much money has been spent says nothing. It is necessary to assess where the project stands in terms of time and then assess where it stands financially in the light of what has been done. The earned value method uses eight basic concepts to help do this:

PV = planned value: the cumulative budget value of what was planned to do until the assessment day

AC = actual cost for what has been done until the assessment day

EV = earned value: the *budget* value of what has been done until assessment day

SV = scheduled variance: the difference between EV and PV; positive SV = ahead of schedule and negative = behind schedule (Note that one must have worked with the activities on the critical path and therefore must also control the CPM time schedule.)

BAC = budgeted cost at completion

CV = cost variance: the difference between EV and AC; a positive value = well placed financially; negative value = over budget

CPI = cost performance index: the ratio between EV and AC

SPI = schedule performance index: the ratio of EV and PV

To assess the earned value—that is, the budget value of performed work—the following assessments are recommended:

- Completed activities as a percentage of BAC
- Measurable milestone completed
- Measurable milestone completed and percentage of BAC
- Technical results—measurable results

However, I prefer the following approach to assess the work performed (variation of percent complete), as shown by the following example. It also gives the person who evaluates the cost a commitment to meet the figures.

Planned value. The PV can be read from the reference plan that is built up according to Figure 6.30. Also, see Section 6.8.3. PV after 5 months is $600,000.

Earned value. The project schedule situation is reviewed after 5 months; the results are shown in Figure 6.31. The line shows that activity A is a little behind schedule. The EV can be obtained by multiplying the performed percentage with the budget at completion for the activity. I have noticed that I get better EV value if I ask the question: "How much will it cost me to finish this activity?" and take this amount from the budget for the activity—for example, assess how well activity A has been done by estimating how much it will cost to complete the task ($25,000) and take this from the budget of the activity ($300,000). Activity B is also a little behind (cost to complete $125,000); however, in activity C, more has been done than was planned to have completed after 5 months. Thus, activity C is ahead of schedule. The cost to complete is estimated to be $25,000.

The EV after 5 months is the difference between the budgeted amount and the estimated remaining cost. In this case, ($300,000 – $25,000) + ($300,000 – $125,000) + ($200,000 – $25,000) = $625,000.

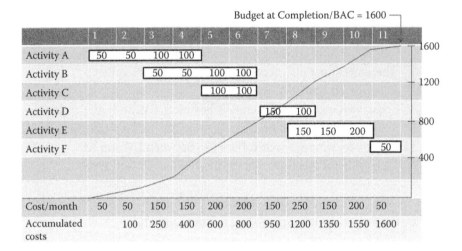

FIGURE 6.30
Planned value and BAC.

Actual cost. It has also been ascertained that the actual cost (AC) after 5 months is $525,000.

Now the EV and the AC are added to the reference plan (Figure 6.32). We have also added the assessment figures after 3 months. EV = $200,000 and AC = $150,000.

The difference between what has been done (budgeted amount), EV, and what should have been done, PV, is the schedule variance, SV. The

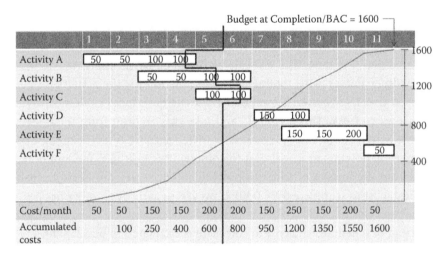

FIGURE 6.31
Reconciliation of the schedule and calculated earned value.

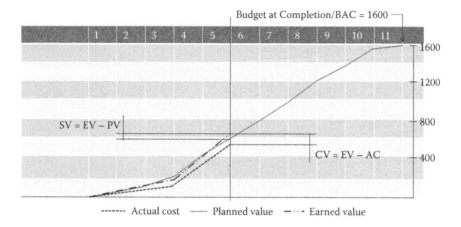

FIGURE 6.32

EV curves after 5 months. Graphic image of the time situation and financial situation.

difference between what has been done (budgeted amount), EV, and what it actually has cost, AC, is the cost variance, CV.

$$SV = EV - PV = 625,000 - 600,000 > 0$$

The project is ahead of schedule, but only if it has been worked on the critical path. **Check the schedule too!**

$$CV = EV - AC = 625,000 - 525,000 > 0 \text{—better than budget}$$

What conclusions can be drawn about the date for completion and the final cost of the project?

- The project can continue to (extrapolate the slope) the way it has been done since the last assessment.
- The project can continue to (extrapolate the slope) the way it has been done since the start.
- The project can continue as planned.
- The project can develop after other assessments.

If the SPI and CPI are used in an intelligent way, the result will give the uncertainty box shown in Figure 6.33. Looking at the figure, which includes an assessment after 9 months (AC = $1,400,000, PV = $1,350,000,

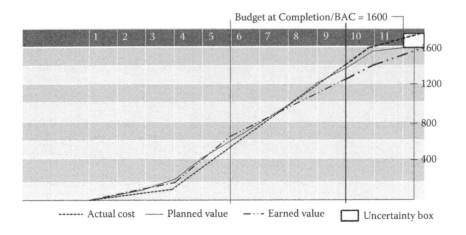

FIGURE 6.33
EV curve after 9 months and an uncertainty box.

and EV = $1,275,000), and extrapolating the curves to the finished project results in an uncertainty box.

The quality of the calculated final cost is based on the assumption that the efficiency is the same as earlier in the project. Are the same activities performed? Remember that if the design is doing better or worse than planned, this does not necessarily mean that the production will go the same way. Divide the project into groups of activities and then follow up with the earned value method.

Experience unfortunately shows that if the design or production goes poorly due to many changes, it will continue in the same way. This means that as soon as signals of this are received, it is necessary to act decisively. The method can be misleading unless critical path activities have been undertaken. Work should always be controlled against the network diagram. In addition, a separate EV diagram can be established for changes and extra payment made to the contractors for unclear drawings, etc.

6.8.5.7 Invoice

No project is controlled by invoice control. But invoice control prevents erroneous payments, including

- Double billing
- Payment of change costs that have not been agreed to

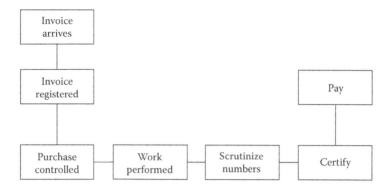

FIGURE 6.34
The invoice process flow.

- Payment for work not yet performed
- No payment until all outstanding issues from the final inspection have been corrected

A company has special payment procedures and accounting requirements for invoice processing. The invoice must be registered, scrutinized, certified, and paid (see Figure 6.34). This takes time. Never expect an invoice to be paid the day after it was sent.

Pay the undisputed parts of an invoice. Deteriorating relations also arise when the contractor's accounting department sends out invoices according to a payment schedule even though the contractor is behind schedule. All invoices should be passed by the TL or CM for approval before they are posted. Discuss with the client if the invoice amount is not in accordance with requirements. This reduces irritation and extra work for both client and contractor.

Given economic uncertainties, the first invoice should not be paid before bonds and proof of insurance have been received.

6.9 RESOURCE MANAGEMENT

WHAT human resources are needed?
WHAT are the needs for expertise?
WHAT equipment and office resources are needed?
WHAT are the responsibilities of various project members?

WHAT is the authority of different project members?

HOW can participants be encouraged to work toward the same goal?

HOW is collaboration facilitated between various participants?

HOW can participants develop the project?

HOW can questioning be addressed and conflicts avoided?

WHEN are the resources needed?

- **Purpose**
  - Ensure that adequate human resources in terms of expertise and capacity are available at the right time (man the project).
  - Ensure that the working climate in the project contributes to an efficient implementation (develop the project team).
  - Ensure that responsibility, authority, and reporting are understood by all.
  - Reduce the likelihood and consequences of conflicts.
  - Ensure that necessary equipment, information systems, temporary workshops, and office space for the staff are available.

6.9.1 General

Project work requires human resources and resources in the form of equipment, information systems, and facilities. The perennial questions are "How many are needed?" "How much is needed? "Where and when are they needed?" The need for special machinery and expertise has to be investigated. A particular resource may be needed in a different place in another project at the same time. Resource planning and scheduling therefore go hand in hand. (Compare Section 6.7.5.)

Lack of resources is a recurring problem for a project. Often this is due to poor resource planning or program management. Having full-time employed members is preferable because there will be no need to fight with the line organization or other projects for resources when they are needed A project team can vary in size and composition during a project. It is one of the benefits of a project. By planning resources, the company may utilize employees more efficiently and thereby become more competitive. For shared resources, it is appropriate, during planning, to make arrangements for temporary resources with internal or external resource owners.

If a WBS has be done, the knowledge gained from it can be used. As usual, there are optimists and pessimists when assessing resources. Consider whether the knowledge in Section 6.11 can be used.

6.9.2 Human Resources, Organization

Project resources' responsibility, authority, and reporting must be made clear in a simple manner. Preferably this is done with a graphical organization chart. Each square corresponds to a role in the project. Many companies have templates for organizational charts and descriptions of responsibility or power that can be adapted to the current project. In small projects, the same person can have many roles. See the organization charts in Figures 2.1–2.3 in Chapter 2.

The planning phase includes identifying and documenting roles, responsibilities, and reporting lines. The need for staff with different qualifications must be planned for. Once this is done, the project must be staffed according to the identified needs. Sometimes it is necessary to educate or hire staff.

The PM leads the project team by providing feedback and following the accomplishments of the participants. Sometimes, staff that do not have the right qualifications or attitude must be relocated. With various activities, the project participants are welded into a project-loyal team. Successful projects are not the result of great resources, but rather of perseverance. An environment where team members feel safe must be created. It must be fun to go to work even when it is tough. Participants come from different organizations and have different training and experience. The PM must understand and support team members.

6.9.2.1 Teambuilding

Henry Ford is quoted as saying,

> *"Coming together is the beginning, keeping together is the progress, working together is success."*

During the Korean War the psychologist Will Schutz studied how team members interact and communicate in a group. The results were presented in 1958 as the FIRO (fundamental interpersonal relationship orientation) theory. This theory explains how humans develop in groups and projects.

How much do group members interact, socialize, and take responsibility? FIRO divides the project into three phases: affection/openness, control, and inclusion. In the affection/openness phase, members of the group orient themselves and decide what they like. In the control phase, they position themselves to create a motivation to influence the group. In the inclusion phase, they act openly and honestly with each other. In a project with various activities, it is necessary to ensure that everyone enters the phases as quickly as possible. That is when the project team is most effective.

Some projects have special requirements for security and secrecy. The members must be educated in the processes that will apply.

The team and the participants should be rewarded appropriately. In Sweden, where "the royal Swedish envy is much stronger than the Royal British Navy," one must think twice before deciding on the rewards. Bear in mind that positive criticism is best given when others listen and negative criticism is done privately.

6.9.3 Staff Relationships

The PM's role is also to handle questioning and conflicts between different participants and departments. Questioning can, if managed correctly, lead to increased creativity and better decision making. If it is not handled in the right way, questioning can evolve into conflicts.

Primarily, the project participants should resolve their conflicts themselves. If this cannot be done, the PM should act. He or she must take the initiative for conflict resolution before the conflict assumes such proportions that it harms the project.

6.9.3.1 Relationship Index

How does one measure relationships? In one of my projects with bad relationships, I studied how communication worked between different project participants, and how it worked with people in the line organization. I named this the *relational index*. I assessed the ability to influence project performance, time, and cost—an *influence index*. With an index scale of 1–5, 5 is a good interaction or high impact on the result.

Activity	Index	Impact on Area	Index
Chatting with each other		Scope	
Called person calls back within 24 hours		Time	
Has a positive attitude toward different issues		Cost	
Can maintain an open dialogue on alternative solutions		Sum	
Shares experiences from previous projects		Index (sum/3)	
Wants to comment on the suggestions			
Eats lunch together with others, whenever possible			
Talk about their families			
Total			
Index (sum/8)			

By placing the index of various persons in a relationship diagram, I was able to identify people who were "time bombs" that could mess up the project really badly. An early action against the time bombs can save the project from several conflicts (see Figure 6.35).

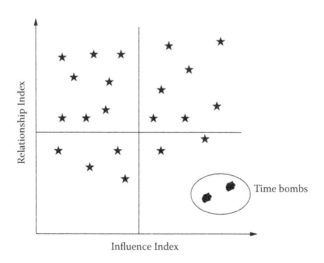

FIGURE 6.35
Identification diagram for relationships and project impact.

6.9.3.2 Challenge and Conflict Management

The PM must be aware that questioning and conflict in projects always occur. The time, money, or resources many times are insufficient for the individual member. Questioning and conflict can also occur due to established territory, culture, informal networks, organizational cultures, or envy. People are different in nature and must be treated with respect. Project participants who fall in the black box in the lower corner in Figure 6.36 are "corpses in the cargo" for the project and have no place in it. People in the upper left corner must be handled by the PM to avoid anarchy.

Always consider very carefully whether to trigger a conflict or not. Prestige is an uncontrollable weapon. Challenge of technology or production methods may develop into conflict for reasons of prestige. If questioning is addressed early and properly, however, it can provide a better solution for the project.

To resolve a conflict, it is necessary to

- Listen actively
- Confirm facts
- Show respect
- Be honest
- Show sincere empathy
- Agree on the issue

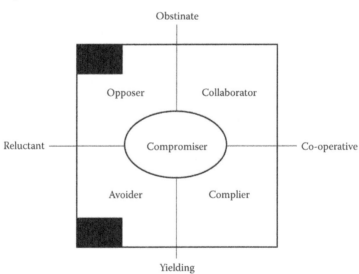

FIGURE 6.36
Identification chart for cooperation and obstinacies.

In conflict resolution, it is necessary to distinguish between cause and person. The process normally follows the following activities:

- Introduce.
- Ascertain the facts.
- Clarify what is going well.
- Clarify what does not work.
- Find solutions to the conflict.
- Confirm the agreement.
- Follow up.

In projects, people with whom others are uncomfortable show up at regular intervals: the insecure person who seeks help and support all the time, the joker that people are not sure when he or she is joking, and the silent partner who does not want to get involved. Some people question everything and others are self-oriented and want everything to be adapted to their plans.

One of the adult world's great secrets, that all men are not equal, must not be unknown to the PM. Those who think they always have to work with difficult people should consider what the lowest common denominator for these projects is.

When it is necessary to criticize a person for behavior, one can start by talking about how people perceive the behavior. Then, one can ask, "How do you think others perceive your behavior? What is behind it?" To start a dialogue, it is necessary to wait for answers. Otherwise, if the monologue continues, it just runs off the person in question. Criticism should be given privately and one should not apologize for criticizing.

6.9.4 Equipment and Computer Systems

What equipment is needed? It can be anything from a giant floating crane to construction scaffolding and special machines. When is it needed and when is it available? What computer programs and software versions should be used? Do all members need computers? If three-dimensional design is at various stages, which stage should be used? Is a central server needed to manage the planned communication and information within the project? Does any part of the project take place abroad? What does this mean in requirements for equipment and information systems?

6.9.5 Premises

What premises are needed for personnel, equipment, manufacturing, and warehousing? If the building site is located in a place where people cannot get home each day, what should be done? Is the project responsible for food and lodging? This is not a small task if the project is located deep in Angola, for example.

Contractors' need of parking spaces must be resolved. When contractor personnel arrive at 6.30 a.m. and grab the client's or users' parking spaces, there will be conflict.

Where can sheds be placed? Are temporary storage areas or parking lots needed?

What are the manufacturing and storage requirements (heating, lighting, ventilation, power supply)?

Not to have meeting rooms can be a big nuisance. Therefore, it is a good habit to book meeting rooms as soon as an agreement on regular meetings is reached.

6.10 COMMUNICATIONS MANAGEMENT

WHAT needs to be communicated and who will have the information?
WHAT needs to be archived?
WHAT are the legal and contractual requirements for information and filing?
HOW and **WHEN** should information about different things be given?
HOW is archiving achieved?

6.10.1 Notable Energy and Dashing Spirit

Good humor and the "devil's embrace" do not help the success of a project. One must have the right information to make the right decisions, produce the right drawings, and manufacture and install properly. Project participants are not the only people who must be informed. Stakeholders who do not have the right information can cause major problems with moves based on rumors.

- **Purpose**
  - Ensure that the right people have the right information at the right time.
  - Ensure that the correct information about the project is archived safely and easily accessible.

The PM should spend time communicating with project participants, stakeholders, and customers. Information can be distributed in meetings, protocols, basic documents, schedules, and databases available online. However, who should receive different information and who should have access to different parts of a project server must be determined.

Some projects or parts of projects require great secrecy. What restrictions are there? How can this be accounted for? What procedures must be followed?

There are many channels of information in a project. Using the formula $n (n - 1)/2$, where n = number of stakeholders, the number of information channels can be calculated. With 10 people, there will be 45 different channels of communication. The dissemination of information must therefore be organized.

Communication is not easy. Users describe their needs. These are handled by architects, technical consultants, purchasers, and the general contractors' and subcontractors' supervisors and are ultimately carried out by a craftsmen who did not receive any information about the original need. Much has been lost along the way. Throughout the project, the PM must ensure that the knowledge of user needs is retained. Within the industry, successive approvals of draft documents, system/basic documentation, construction documents, and shop drawings are used. Special "technology" meetings (see Section 6.2.2 for meetings during the production phase) can also help manufacturers and craftsmen to understand why the plan is drawn and described in a certain way.

Communication is difficult. Communications are often disturbed by unclear messages and difficulty in understanding them. The basic model can be described as in Figure 6.37.

If the information is written, choice of words and allusions must especially be considered. Which media should be chosen? The choice can be between protocols, memos, formal reports, e-mail, etc.

For an oral presentation, appropriate visual aids must be chosen and body language considered. It is not just the spoken words that convey a message. Metalanguage, emphasis, tone, volume, rhythm, and speed affect

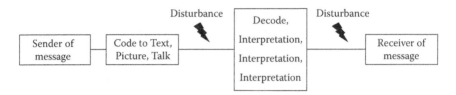

FIGURE 6.37
Graphical picture of communication.

the information. We also send messages with kinetic language, facial expression, head position, eyes, mouth, gestures, posture, etc.

Information is also received in different ways:

- Informed, but have not listened
- Listened, but have not understood
- Understood, but have not accepted
- Accepted, but have not implemented
- Implemented, but only briefly

When a presentation begins, the message should be formulated as assertions or questions. A few key words should be selected and then used to do background description. Describe what is desired to happen. In order to get this accepted, use logical arguments, appeals to emotions, and, above all, confidence in the task. Conclude by describing the case as a common goal.

WHAT, HOW, WHEN, WHY, and by WHOM are questions to be answered.

6.10.2 Information about Changes

Documentation and dissemination of changes are important to file (see Section 6.4.2). During the design phase, "internal changes" (within the project scope) are usually decided at PMD meetings and information disseminated via the protocol/minutes from meetings.

Revisions of bid documents and building documents should not proceed without PM approval. The PM determines, after consultation with the DM and construction manager, whether the revision should be performed or not. After bid documents, any change in drawing or in a document should be accompanied by a revision memo. This should include information about which pages, drawings, or items have been changed and a brief description of the revision.

In complex projects, the contractor and site management regularly discover coordination deficiencies in documents. The work requires quick fixes to avoid delays. If a consultant discovers a "minor" revision, the SM should be informed. The SM determines whether the revision should be sent out immediately or together with another revision. The construction manager then determines whether to inform the contractor immediately or wait until the next revision memo. The PM or a person designated by him or her should be informed of any revisions even if they are just "clarifications."

Written confirmation of the submitted revisions must be obtained from contractors.

- **Responsibility**
  - The TL is responsible for ensuring that the revision process is not started until the revision application is approved.
  - The SM is responsible for ensuring that the contractor acknowledges the delivered information about changes.

6.10.3 Information Exchanges

The choice and range of the information exchange depends on the size and length of the project, available communication facilities, information language, and stakeholders' requirements. Written information takes the form of

- Form for questions and answers
- Inspector memos
- Contractor memos
- Letters
- Protocols
- Incident reports
- Sketches
- Drawings
- Descriptions

The distribution can be by mail, fax, e-mail, or courier.

All written documentation must be dated and have information about the project, building, senders, and recipients (including recipients who receive information only for information purposes). Distribution lists are established for the distribution of documents during the design phase and production phase. Project information is normally stored directly into a shared server, if available. In order to manage and store distributed e-mails effectively, there should be process or routine. This is especially important with multiple builders. The e-mail header should always note project name, or contract number and project name.

6.10.3.1 Communication with Contractors

All communications with contractors should be done through the form for questions and answers. When drawing change notifications are transmitted, the communication should be documented in writing. Forms for questions and answers should be dated and sequentially numbered.

6.10.4 Information Exchanges

Progress reports are made at times that are appropriate for the project. Examples of timing for progress reports include

- Feasibility study
- Schematic/basic design
- Important procurements
- Every 4–6 weeks during regular production (shorter periods mean "garbage in/garbage out")
- Final inspection
- Project close out

Contractors should do status reporting to the client each month and when something special occurs. More frequent progress reports steal time and do not give the client a better base for decisions. In special situations, the client should have the right to request a quick presentation of the implemented activities and results.

The progress reports should be concise and include the following:

- Descriptions that focus on the remaining and future tasks—what to do
- Descriptions of what has not been done that should have been done and of the action needed to get back on track
- Any changes in the functional program and/or the construction programs
- Time

- Economics
- Special measures to compensate for lost time and what to do to get back within the budget framework
- Risk and critical items

6.10.5 Documentation

How will the documents be filed and archived during and after project delivery? What are the requirements for the documents that will be delivered for operation and maintenance?

6.10.5.1 Validation and Compliance

The documentation of the validation work will be done both at the design stage and during the production phase. The requirements are normally specified in the SOPs (see Section 6.6.6).

6.10.5.2 Security for Document Management during the Design Phase

The consultants are responsible for backing up digital documents. This should be reflected in the contract between client and consultant. Procedures for safekeeping other documents are normally shown in quality plans.

6.10.5.3 Document Management during the Production Phase

- **Purpose**
  - Ensure documentation of actual performance.

Contractors should include information on the existing drawings. When the job has been completed all these changes are transferred to the as-built documents.

The basis for the as-built documents should include two sets of drawings. The set stored at the site is where the current changes will be incorporated, at least once per week. The second set of drawings should be the "last" as-built drawings and compile all the changes after the contract work is completed. These drawings should be sent to the final inspectors (e.g., 2 weeks before the final inspection). The inspectors can then control the drawings before and under the final inspection. This is a simple requirement that should be entered into the contract.

6.10.5.4 Document Management after Final Inspection

- **Purpose**
  - Ensure that current information on buildings, installations, equipment, and furnishings is easily available.

As-built documentation should be prepared and filed before the project is completed. This documentation includes descriptions. There must be agreed policies for digital documents; line thicknesses, colors, layers, etc., must be established. In addition, whether the client intends to use a special program for his or her own as-built drawing should be stated. This is especially important if the contractor should deliver the as-built documents.

Maintenance instructions and lists of products will be provided not only by installation contractors but also by the builder. Think of all the hardware that is installed in doors, windows, and kitchens. Where can a new one be found when the old one is broken?

6.10.6 Archiving

- **Purpose**
  - Ensure that documents are protected so that they meet authority and code requirements for filing.
  - Ensure that documents of permanent economic value (such as land purchases) are stored in a fireproof location.
  - Ensure that documents of limited economic value (such as contracts, agreements) are stored in a fireproof location during the warranty period.
  - Ensure that documents with confidentiality requirements (e.g., bids) are filed in a fireproof location with limited availability.
  - Ensure that documents are stored according to a specific system so that they can be found easily by someone other than the person who filed them.
  - Ensure that the PM and other employees know where and to whom documents must be delivered for filing.
  - Ensure that the project's filing requirements with respect to validation and compliance are met.
  - Ensure that bonds, guarantees, and insurance documents are protected against theft and fire.

6.11 RISK AND UNCERTAINTY MANAGEMENT

WHAT can go wrong in the project?
WHAT can go better than planned?
WHAT things are indistinct and uncertain?
HOW can the risks that threaten to cause the project to fail be reduced?
HOW can the opportunities to do better than planned be increased?
HOW can the risks and uncertainties that must be accepted be covered?

6.11.1 Introduction to Risk Management

In all projects, there is indistinctness, which means that assumptions must be made. These assumptions are often uncertain and can involve time, cost, scope, or resources. Sometimes an event affects the project. Political decisions, strikes, exchange rates, and organization changes can change the project's prerequisites. Changed prerequisites can also be caused by the PM, someone else, or nature.

Events can affect the project negatively (threats) or positively (opportunities). The consequences, if the event occurs, can be small or large, and they may even be a catastrophe for the project.

There are business risks resulting in gain or loss. Pure risks and "insurance risks" only give losses. These are normally insured if the effect is unacceptable for the project.

Some events can be foreseen; these are known as identified risks (known unknowns), but the probability of their occurrence is uncertain. Other events have not been identified and are therefore unknown risks (unknown unknowns). In February 2002, before the Iraq war, US Secretary of Defense Donald Rumsfeld discussed risks at a Department of Defense news briefing:

> Reports that say that something hasn't happened are always interesting to me because, as we know, there are known knowns; there are things we know we know. We also know there are known unknowns; that is to say, we know there are some things we do not know. But there are also unknown unknowns—the ones we don't know we don't know. And if one looks throughout the history of our country and other free countries, it is the latter category that tend to be the difficult ones.

This statement won the 2003 Foot in Mouth award from the Plain English Campaign and was also hailed as an example of found poetry by Hart Seely in *Slate* magazine. However, a person with relatively little knowledge of risk management might understand what he says.

The risk management process shows how to deal with the uncertainties for the success of projects. Laws and regulations in most countries require that developers, consultants, suppliers, and contractors identify environmental, safety, and health risks and threats (see Sections 6.1.4 and 6.5).

Following laws and regulations is not enough for a project to succeed. The project requires a thorough risk assessment of site location, contract work, time, finances, project organization, and selected or proposed construction solutions and methods. Once an *event* or *uncertainty* has been identified, the *consequence* and the *likelihood* that it will occur must be described and assessed. Perhaps "triggers" can be identified. When the probability (P) and impact (I) are assessed, the risk exposure can be obtained: $R_x = P \times I$. This is illustrated in Figure 6.38.

Many risk management techniques and tools are on the market. Most of them are named after the method or tool they use, but all are based on the same process: identify, analyze, prioritize, plan risk response, act,

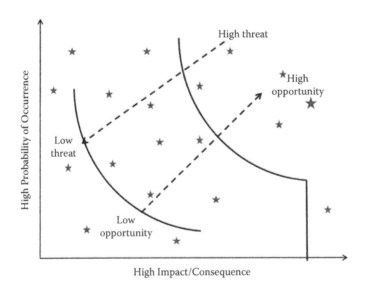

FIGURE 6.38
Risk exposure depending on the probability and consequence.

evaluate, and document. Common methods or tools are RPN (risk priority number), FMEA (failure modes and effects analysis), HAZOP (hazard operational analysis), FTA (fault tree analyzer), fishbone diagram, what if, risk matrix, and mini risk.

Many times it is sufficient to use the mini risk method, where the key is to identify and gain consensus about future threats and opportunities. This method analyzes the likelihood and impact of quality. The highest risk exposure figure ($R_x = P \times I$) gives the highest priority. Of course, more precise numerical values are more interesting for budgets and bidding. Here the expected value or the Lichtenberg method can be used.

6.11.2 Introduction to Risk Management

6.11.2.1 Risk Perception

Risk events are circumstance dependent. No textbook solutions or templates include all risks. The project team must use the tools and methods in a rational way. A risk event can influence or trigger other risk events. Risks are also valued differently. Events that are close in time are often valued to have greater consequences than those that are further ahead. In fact, however, it is often the opposite. Although at the beginning of a project, there are many uncertainties, there is also time to correct an impact. Later in the project, time is perhaps the key to cope with when a consequence occurs.

The perception of risk is influenced by information. Is it inadequate, unreliable, or prepared under time constraints? Different people perceive risks differently. Stamp collectors (risk avoiders) and skydivers (risk takers) value the consequences of an event differently. Men and younger people are willing to take more risks than women and the elderly. This applies generally, not for each project. I had in one of my classes in risk management a woman who had climbed Mount Everest and a 60-year-old free-diver.

In the analysis phase, it is important for the PM to know what kinds of people are doing the valuation. The cultures of different companies and countries have different positions in the valuation of consequences and what can be accepted. People's perceptions of risk and their approaches are different.

Leo Tolstoy wrote in *War and Peace,*

> In critical situations, there are two equally strong voices that speak within us. The one that you should make clear what the dangers are and figure out how to deal with it. The second that you should seek to forget the risk until it cannot be avoided and instead think of anything nicer.

6.11.3 The Risk Management Process

When a risk event occurs, it is not always due to any single factor; it can be difficult to distinguish between cause and effect when a problem occurs. Therefore, it is necessary to work with a structured process. In the modern project management process, systematic risk management has been implemented. This means that uncertainties are studied in a systematic way (e.g., the ESI™ way):

- Plan the risk management work in the project.
- Identify.
- Analyze.
- Prioritize.
- Plan risk response (actions for the priority risks, threats, and opportunities).
- Plan actions to make threats acceptable and raise the chance of opportunities happening.
- Implement actions and monitor changing situations and triggers (early warning system).
- Assess and document experiences.

See Figure 6.39.

If one does not accept the impact, then one must act. Since the time and money to deal with all threats are not available, one only deals with the prioritized ones. Risk response is planned with the following strategies (Figure 6.40):

- **Avoid** the risk; for example, eliminate the cause or do not perform the activity.
- **Transfer** the risk to someone else—for example, bonds, guarantees, or subcontractors.
- **Mitigate** and reduce the likelihood of the risk occurring and/or impact if it does occur.

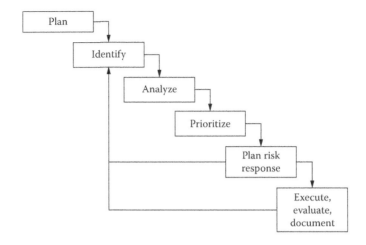

FIGURE 6.39
The risk management process.

In the same way, the opportunities are addressed:

- **Exploit** the opportunity.
- **Enhance** and increase the likelihood and/or consequence.
- **Share** the opportunity with the client or third parties.

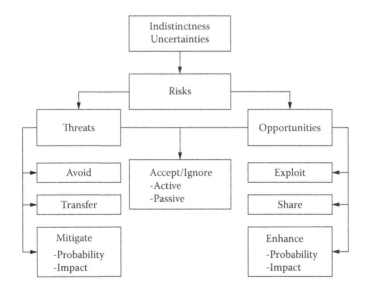

FIGURE 6.40
Response strategies to threats and opportunities.

For the accepted threats—those for which risk response is not planned, there is a choice of

- **Active** acceptance—nothing is done now, but a plan is in place that will mitigate the damage if the event occurs
- **Passive** acceptance

The information is compiled into a risk management plan (RMP) with the response actions and the name of the person who is responsible for effecting them. The RMP is monitored at appropriate regular meetings during the project.

6.11.4 Risks in Projects

It is necessary to distinguish between the uncertainties of the future product and uncertainties in the project. The laws and codes on the environment and the work environment require that the safety, health, and environment (SHE) in the project and product be taken into account. The project knowledge and assumptions must be transferred to both operating personnel and to the user of the product (Figure 6.41).

During the design, the threats and opportunities that affect the product must be considered. A threat may be that a ship can collide with a bridge pier. One opportunity is to study how plants and animal life develop on a newly built artificial island. This was done, for example, on Pepparholm, an artificial island in the Öresund between Sweden and Denmark. Another opportunity could have been not to demolish the temporary port on the island and instead build a marina and a profitable casino.

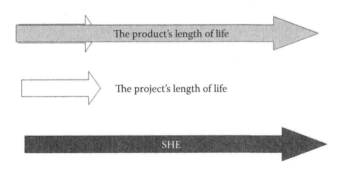

FIGURE 6.41
Project risks.

Other uncertainties in the construction business are wind forces, material grade, vehicle impact, etc. These issues have different organizations and legal bodies that can help. They have specified codes and design guidelines, allowable stresses, etc. This does not mean that the project members can stop thinking for themselves. The guidelines are simply guidelines and minimum requirements.

For a project, the threats and opportunities are seen. A method of production is not chosen based only on the cost and time. Also considered are the uncertainty and whether the project will be ready on time, within budget, and with the right quality. Is there a risk that customers and suppliers can go bankrupt or that skilled employees will move to another company during the project?

Both buyers and suppliers experience uncertainty in pricing. How will economic uncertainty be divided between the parties at contract signing? Can risks be identified and how to settle them be agreed upon rather than simply shifting the risk to the other party? The other party may charge for taking the risks. If the project is FULLY COVERED against uncertainties "in order to sleep well at night," it may be more expensive than if a decision is made about how to settle the uncertainty. All threats do not occur.

Common risk-related events are

- Uncertainties in the basic material for the feasibility study
- Disruptions of ongoing activities
- Poorly defined tasks
- Missed information
- Short time frames
- Late decisions and permits
- Unsafe or difficult foundation conditions
- Weather conditions at the "wrong" time of the year for construction
- Temperatures below 0°C (32°F)
- Product defects or incorrect performance
- Poor dehydration or drying-out of concrete
- Heavy loads from material and equipment on soil that will collapse
- Damaged goods or late deliveries to the site
- Ignorance or carelessness due to the human factor
- The money situations of clients, suppliers, or contractors
- Currency fluctuations
- Water and power cuts

- Fire
- Water damage during construction from the roof openings, windows, etc.

6.11.5 Risk Management—A Full Project-Life Activity

Risk management work is an ongoing process throughout the project's life. Identification and analysis are carried out gradually as knowledge increases. Therefore, the whole project, a particular phase, or a specific task or inspection test is studied. The risk management process should take place

- In connection with the feasibility study
- In connection with the system and basic design
- In connection with the design start
- Prior to contract negotiations
- At the start of production
- For machinery, equipment, furnishings
- In ad hoc situations that may affect the project

Risk management should involve

- Environment, safety, health, and security
- The design work
- Production in the factories and on the site
- Deliveries
- Time
- Economics
- Unclear objective or contract texts
- Organizational weaknesses
- Authorities
- Legal questions

Let common sense rule when risk management is performed. One project threat, of course, is that there is too little common sense in the project and bureaucratic rules govern.

6.11.5.1 Comprehensive Risk Analysis for Projects

A project often starts by evaluating the project risk on a risk scale. This provides guidance for in-depth analysis and knowledge of management requirements for the project and the management-level for implementing decisions. By giving different points to size, complexity, number of suppliers, etc., the project sum will help to make a decision. Examples of values to analyze include:

Project value	<$100,000 0 points	$0.1–1 mil. 20 points	$1–10 mil 40 points	>$10 mil 80 points	
The working volume in man-months (mm)	<25 mm 0 points	25–50 mm 20 points	50–150 mm 40 points	>150 mm 80 points	
Number of contracts/ subcontracts	<5 0 points	6–10 20 points	11–15 40 points	>16 80 points	
Problems with client and users	Little 0 points	Normal 10 points	Difficult 30 points		
New or existing technology	Well known and done before 0 points	Known but no personal experience 20 points	Unknown[a] 80 points		
Time	Well balanced time with buffer 0 points	Tough schedule 20 points	"Impossible" schedule 40 points		
				Total sum	

[a] Requires in-depth analysis.

Sometimes the sum is multiplied by 0.75 when a client is well known and there are good relations; 1.0 is used for an unknown client and 1.25 for a known client with past conflicts. If a certain total sum is exceeded, the project must be studied more deeply and a decision about the project made on a high level.

6.11.6 Practical Risk Management

The PM decides when and in what areas a risk management process should be implemented. Risk management work (identification, analysis/prioritization, planned risk response, implementation, and documentation) is conducted by the project staff. This is not a one-man job. The PM decides

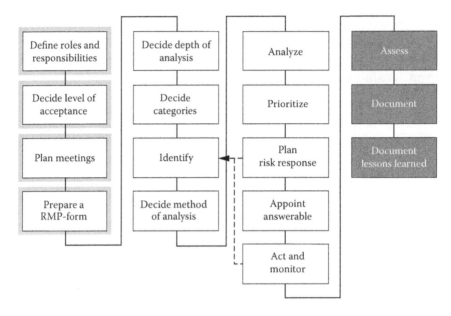

FIGURE 6.42
The process of risk management work in a project.

the methods, tools, and acceptance levels in accordance with company or client policies (see Figure 6.42):

- **Start-up**
  - Determine the methods for identification and analysis.
  - Define roles and responsibilities.
  - Prepare an RMP template.
  - Decide level of acceptance.
  - Identify categories.
  - Call for an identification meeting.
- **Implementation**
  - Risk identification: Identify risk events for the project. The identification should take into account not only the risks to the money situation, time and quality, environment, and technology, but also risks in resources, logistics, coordination, safety, reliability, responsibility, personal conflicts, and information. At risk identification, a list with experience from previous projects should complement an identification method.

- Analyze and prioritize: Evaluate the likelihood (probability value) for the risk events to happen and the impact (impact value) if they happen. List unacceptable events.
- Plan risk response: Develop plans to address risks (threats and opportunities) for prioritization. Once a risk has become a high priority, it requires a plan of action to lower the threat to an acceptable level or to increase the possibility to a level where it is likely to happen. It is also necessary to verify that this action does not cause new, unwanted effects. Designate a person to be responsible for implementing the measures.
- Act, monitor, evaluate, and document: Perform actions in accordance with action plans. Action plans should always be updated so that they are ready for implementation. Follow up the RMP at appropriate meetings. The management of risk response also includes information to stakeholders, the listing of any new events, and documentation.
- **Finish**
  - Continuously evaluate the impact of measures during the project. Did the event happen? What was the impact? What lessons were learned?
  - Document continuously during the project and compile at the end of the project.
  - Share and document lessons learned.

6.11.6.1 Risk Management Plan

The information about the risk management process is collected in an RMP (see Appendix K). The layout of an RMP will vary depending on what is needed. The following headings are recommended:

- Identification number (e.g., WBS codes and/or category)
- Event description
- Description of consequence or impact
- Probability value
- Impact value
- Risk exposure
- Priority
- Risk response
- Dates
- Answerable

Sometimes there are also columns for

- Triggers, EWS (early warning system)
- Evaluation of probability, consequence, and risk exposure if risk response is performed
- Cost of risk response
- Contingency plan
- Contingency reserves

The RMP provides a logical basis to explain how one has chosen to act regarding uncertainty issues. It is used during the implementation phase to ensure that what has been decided is done and that nothing is forgotten.

It is important that stakeholders understand what the uncertainties are and what the risk response means. RMP is used to communicate with stakeholders. It helps to explain what has been taken into account and the steps taken to minimize threats and maximize opportunities.

6.11.7 Risk Identification

Risk identification is not a one-man job. The identification is done in groups with both experienced and less experienced members. Often, it is good idea to invite someone who has been in a similar project. Risks related to both the project and the product should be identified and managed. The identification should take into account not only the risks to the money situation, time and quality, environment, and technology, but also risks in resources, logistics, coordination, safety, reliability, responsibility, personal conflicts, information, organizational risks, customer relationships, etc.

A very common error that causes delays and cost is that, during the identification phase, analysis of the event begins and planning the risk response starts. There will never be enough time or money to plan risk response for all events. That is why the uncertainties are analyzed and prioritized later in the process.

Before the identification process begins, the PM must decide what should likely be studied. Is it the entire project or, for example, just the design phase or work underwater? The PM must also decide which method of identification should be used and which information and assumptions apply.

There are constantly evolving new situations. If the excavated material is put on an inappropriate spot, for example, the soil pressure can be too much and a sheet-pile collapse will result.

6.11.7.1 Event and Impact Description

To describe an uncertain event or its consequence clearly is a difficult task in the identifying work. Often the descriptions are too general: "Unless we get the drawings in time, the project will be delayed." The event and the consequence are, of course, accurately described, but the "root of evil" must be found. Why are the drawings delayed? Loss of decisions, changes, etc.? The risk event must therefore be reworded. For example, **if** the system/basic design drawings have not been approved before dd/mm/yy, the construction documents cannot be delivered as scheduled and the project will be delayed. Unless the architect delivers basic drawings by dd-mm-yy, the mechanical documents cannot be delivered as scheduled. This will cause delays. **If** a watertight house is not created before dd/mm/yy, there is a high risk that rain will enter the building and later cause mold damage. One trick that facilitates the description of events and impacts is to begin with the word **if.**

Many RMP templates have one column for the event description and impact statement. It is often easier if there are two columns. Sometimes, an event is identified but the consequences cannot be described because there can be more than one. For example, if, during the rebuilding process, a fire in a culvert crossing will result in electric wiring damage, what is the electric wiring supplying? It may be necessary to contact the operational and maintenance department or an electrical consultant to get an answer.

6.11.7.2 Categories

To facilitate the identification process, it often helps to identify various categories and discuss each category separately. There are *internal* (within team control) and *external* (beyond team control) *categories.*

Examples of external categories include

- Political decisions
- Taxes
- Currency fluctuations
- Weather

- Natural disasters
- New owners, new management, etc.
- IT changes
- Strikes
- Nationalization

Examples of internal categories include

- Technology
- Installation
- Law
- Economics
- Time
- Resources
- Work, stress, illness
- Environment
- Fire
- IT
- Sabotage or deception
- Bankruptcies
- Quality

6.11.7.3 Brainstorming Process

A moderator of a group of 6–10 persons is appointed to chair the meeting. Sometimes there can be more than one group, but each group should not be larger than 10–12 people. Participants are encouraged to use other people's ideas to help identify other events. All ideas are written down and no assessments of the proposals are allowed because, if a proposal is judged, some participants' imaginations may be inhibited. All events are compiled in the RMP.

The method is interactive and gives the group synergy but requires a strong moderator. The method is intense, but may be dominated by outspoken individuals. The somewhat quieter person with good ideas may not try to or may not get the time to comment. The moderator must therefore be active and draw everyone into the discussion.

6.11.7.4 *Crawford Slip*

Several variations of the Crawford slip method are available. The method asks the participants to write down their events (e.g., on post-it notes) in silence—one event on each slip. In this way, even the quieter persons can present their events. The events are read out in order to achieve group synergies. One note at a time from each participant can be read or the moderator can compile all the slips (e.g., 10 slips from each person) and then read the events out. Sometimes it can be a good idea to introduce the company risk checklists, if available, at this time. Participants are then invited to write more events, if they can. The use of checklists earlier can limit innovative thinking. The events are compiled in the RMP. The flow of one of my Crawford slips is seen in Figure 6.43.

The method is very effective in obtaining much information in a short time. Time is not lost on the analysis and risk response that will come in the next stage. Information is also on paper. The large amount of information means more work for the moderator, of course.

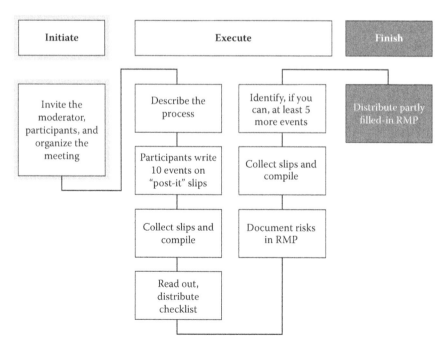

FIGURE 6.43
Flow for Crawford slip.

6.11.7.5 Risk Breakdown Structure

A risk breakdown structure (RBS; see Figure 6.44), which is similar to work with WBS, can be implemented in the identification phase. First, identify categories and then ask **WHAT** can go wrong. Then the questions arise: **HOW** and **WHEN** can it go wrong? **WHAT** are the consequences? This information is compiled in a "risk package" that describes the consistency of time, money, relationships, quality, etc. The information is compiled in the RMP.

6.11.7.6 Fishbone Diagram

The fishbone diagram is an effect–possible causes diagram that is also called an Ishikawa diagram (see Figure 6.45). First, a number of unacceptable effects or adverse events are identified. If a job is taking place near a hospital or a subway station, an unacceptable effect, for example, would be cutting the power to the hospital or subway. Then, a number of categories, such as technical problems, human error, unclear requirements or prerequisites, installation problems, information problems, or others, are identified. Once this is done, events in each category that can cause the power failure are identified.

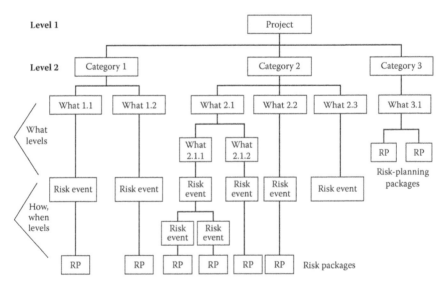

FIGURE 6.44
Risk breakdown structure (RBS). Compare WBS.

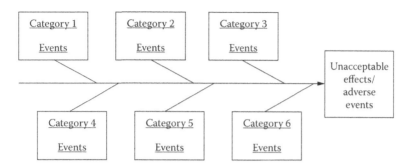

FIGURE 6.45
Fishbone diagram.

6.11.7.7 Other Identification Methods

Sometimes, when faced with unknown issues, experts can be asked for help (the so-called expert method). If the project participants are geographically dispersed, the Delphi method can be used. With the NGT (nominal group technique) method, each participant lists and ranks the 10 greatest risks. In the analogy process, the RMP and/or knowledge from a similar project will be studied, but the focus must be on defining events in the parts where the projects are different.

6.11.8 Risk Analysis

In the analysis phase, the value of the probability and consequence is determined. The risk exposure is calculated as $R_x = P \times I$ or presented graphically in a matrix (see Figure 6.38). It is also necessary to assess whether the risk is recurrent (i.e., if there is a periodicity/frequency, F). The risk exposure is then $R_x = P \times I \times F$.

There are proven tools and techniques for risk analysis:

- Expert judgment
- Risk priority number (RPN)
- Risk matrix
- Expected monetary value
- Statistical sums/PERT or Lichtenberg method
- Computer simulation or Monte Carlo model
- Decision tree

Probable	High	Very high	←— 95%
		High	←— 75%
	Medium	Medium+	←— 60%
		Medium–	←— 40%
Not probable	Low	Low	←— 25%
		Very low	←— 10%

FIGURE 6.46
Bridging key for qualitative and quantitative model. Note that it is the PM who determines the size of the percentage figures if the company's project manual does not specify this.

During the analysis, the probability and impact are assessed. They can be quantified with dollars and hours or qualitatively with colors, adjectives, or numbers (RPN). The quantitative model needs statistics, calculations, etc., and is used primarily in bidding, forecasts, projections, and economic assessments.

The qualitative model requires common sense and a common assessment basis. It is used primarily to identify and prioritize the key risks.

While the identification was made with a mixed group, the analysis process is performed by individuals with expertise in the category. Note that if a qualitative model is used, it is important that all those who analyze use the same scale when assessing. Sometimes one model is used for probability and another for impact. If different methods are used, a "bridging key" may be needed (Figure 6.46).

6.11.8.1 Risk Priority Number

The RPN is a very common method of analysis—a qualitative approach that, on a scale (e.g., from 1 to 5 or from 1 to 10), determines probability that an event occurs and impact, if the event occurs. The risk exposure is the product of probability and impact.

In my opinion, the minimum acceptable scale is 1–4. An even number on the scale will force participants to a deeper analysis. My experience is that when a mini risk (1–3 scale) is used, everything ends up with a

2. Before the assessment, everyone must understand the scale and there needs to be a joint discussion of what a 4 means on a 1–5 scale. Remember that it is the PM's responsibility to define the scale or discuss with company management a directive that fits the project. Another important consequence also must be evaluated: goodwill and negative goodwill.

Based on the risk exposure, the PM should determine the events for which risk response must be planned. For example, when

- $R_x = P \times I > 10$
- $R_x = P \times (I_{time} + I_{cost}) > 18$

If two R_x have the same number, the event with the higher I is always prioritized higher. Events, probability, impact, and risk exposure are documented in an RMP.

6.11.8.2 Risk Matrix

With a qualitative approach on a scale (e.g., from 1 to 5 or with the words low, medium –, medium +, and high), the probability that an event will occur and the impact if the event occurs can be estimated. The valuation is put into a matrix, as shown in Figure 6.47.

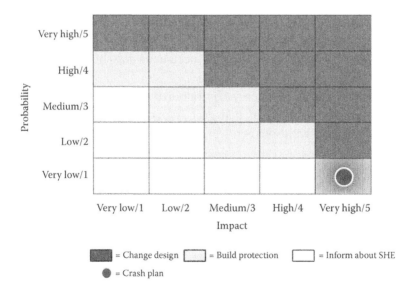

FIGURE 6.47
Risk matrix.

Event	Probability	Impact Value		Expected Monetary Value	
		+	–	+	–
No competition in precast deliveries	15%	$800 000:-		$120 000:-	
Less rock-blasting	30%		$120 000:-		$36 000:-
Snow in March	60%	55 000:-		$33 000:-	
Total		$855 000:-	$120 000:-	$153 000:-	$36 000:- → $117 000:-

FIGURE 6.48
Examples of expected monetary value.

It is normal for the project participants to discuss collectively which fields will be black, gray, or white before the evaluation starts. Of course, the PM can determine this, but the discussion will provide an understanding of what the different colors mean.

6.11.8.3 Uncertainty Assessment with Expected Monetary Value

In economic uncertainty, "expected monetary value," a method that indicates the worst, best, and likely outcomes, can be used. For example, in a project with an estimated cost of $61,000,000, three uncertainties have been identified: no competition in precast deliveries (additional cost estimated at $800,000), less rock blasting than adopted ($120,000), and an unexpected snowstorm in March (additional cost estimated at $55,000). The probability of this occurring has been estimated in Figure 6.48.

Best outcome will be if only the opportunities occur—in this case, $120,000 cheaper than estimated cost. Worst outcome will be if only threats occur—an additional cost of $855,000 ($800,000 + $55,000). The probable expected monetary value is obtained if probability × impact is used for all events ($153,000 – $36,000 = $117,000). A probable cost for the project will be $61,117,000 if the three risks are considered. This is a value to consider when the contingency reserves are discussed.

6.11.8.4 Uncertainty Assessment with PERT

When assessing costs and times, statistical sums can be used. I have found that the method also is useful in early assessment of hole sizes, c/c distance between power-line poles, temperatures, etc.

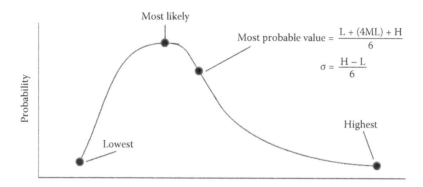

FIGURE 6.49
Examples of PERT curve.

To use the method, a group of 5–10 members is needed. Each member estimates the minimum value (cheapest, fastest, smallest), the most likely value, and the highest value (most expensive, slowest, largest). The method takes into account all the views of the participants, as well as the values from an uncooperative person who thinks that only his assessment is the right one. This person is not ignored and his value should be included; it may be psychologically important in teamwork. Using the PERT method, the spread is assumed to look like that in Figure 6.49. We use the weighted value when the budget or schedule is done. If the standard deviation and variance are used, the probability of different figures can be known. Lowest and highest value is the lowest and highest value that any group member has. Most probable value is the mean value of all the most likely values. How to use the method of projections is shown in Section 6.7.6.

6.11.8.5 Decision Tree

If there are different choices of manufacturing processes and uncertainty must be taken into account, a decision tree model can be used (Figure 6.50).

6.11.9 Priority

There is no time or money to do risk response for all risk events. After the analysis phase, prioritize the risks. The PM decides the ones for which risk response should be planned. Threats and opportunities are prioritized separately.

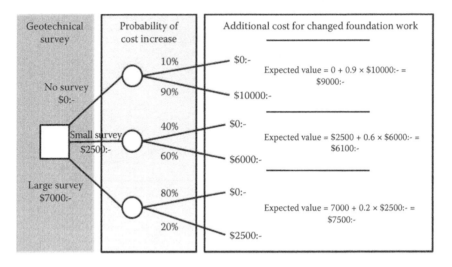

FIGURE 6.50
Decision tree.

6.11.9.1 Risk Priority Number

If the RPN is used, the PM can, for instance, decide that a risk response should be planned for all events with a risk exposure greater than or equal to 10 or 12 (5-point scale). All events that have the highest impact assessment, such as 5 on a 5-point scale or 7 on a 7-point scale, should be crash-planned. There should at least be a contingency plan or emergency plan for these events. When the result is the same R_x figure, the event with high impact gets higher priority than an event with high probability. Events in the range of 6–10 should be monitored a little more during the implementation phase. The consequences or probabilities may change to a higher value and therefore qualify for a risk response.

6.11.9.2 Other Qualitative Analyses

If the analyses have used colors (red, orange, yellow, green, white) or adjectives (very high, VH; high, H; medium, M; low, L; very low, VL), a matrix is created. Together with the team, the PM decides for which boxes in the matrix to plan risk response. It is normal for project participants jointly to discuss the priority fields to be black, gray, or white before the detection starts. Of course, the PM can determine this himself or herself, but the discussion also provides an understanding of what the different colors mean.

6.11.9.3 *Filtering and Comparative Risk Ranking*

Sometimes there is not time to analyze very deeply. Still, it is necessary to assess the need for risk response. In a group of four to nine people, each individual marks his or her own list of priority risks. Then the risks are plotted against each other in a matrix (see Figure 6.51).

Before the comparative risk ranking (CRR) is implemented, it is necessary to **filter** out a number of risks so that the CRR tool can be manageable. The following questions can be asked:

- Is the cost less than $10,000? Do not include.
- Is the delay less than 5 days? Do not include.
- Will the client or project owner notice it? If not, do not include.
- Is the event more than 2 months into the future and will it be included in an ordinary course of analysis? Do not include.
- Has this type of event occurred in any project that anyone has heard of? If not, do not include.
- Can the project team influence the probability or consequence? If not, do not include.

The PM puts two risks against each other and the participants show by hands which risk they have rated higher than the other. The method is

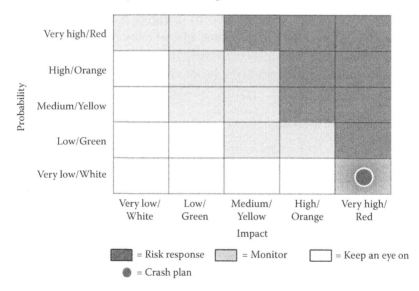

FIGURE 6.51
Priority matrix.

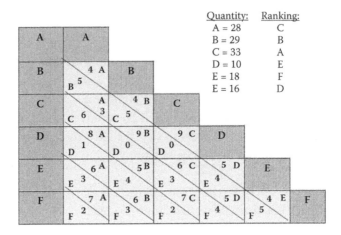

FIGURE 6.52

Comparative risk ranking (CRR) example. In a group of nine persons, four persons thought that A was a greater risk than B and five persons thought the opposite.

illustrated by the following example: Six risks, A, B, C, D, E, and F, have been filtered out and need to be prioritized. The nine persons on the priority team have ranked the risk exposure. In this example, four persons believe that the risk exposure of event A is greater than the risk for event B. Five people think that the risk of B is greater than that of A. The numbers are entered into a matrix as shown in Figure 6.52. A is then tested against C, etc. Then, risk B is compared against risks C, D, E, etc., until all risks have been tested against each other. The results are written into the matrix. By counting the votes of the various risks, a priority is determined.

6.11.10 Risk Response

"You can either take action, or you can lay back and hope for a miracle. Miracles are great, but they are so unpredictable."

—Peter F. Drucker

The list of priority risks is the input parameter for risk response (RR), which is planned for threats and opportunities separately. First, it is necessary to find out who owns the risk and what the client's organization can accept. There may be financial risks, health risks, goodwill or negative goodwill (e.g., child labor), or environmental damage. Begin by asking a few questions:

- **HOW** can the threat be **avoided?**
- **WHAT** can the customer **accept?**
- **HOW** can the probability and/or impact be **reduced?**
- Can the threat be **transferred** to someone else?
- Can this opportunity be **maximized?**

The RR must be consistent with the overall project objectives. RR can interact in strange ways. One action may affect other risks positively or negatively. It is not enough to plan one response for each risk. The proposed actions can be entered into a matrix with the risks at the side and the responses at the top. Each RR is marked as positive or negative. Does it influence more than one risk? Analyze the result and pick one RR. When an action is selected, it is necessary to ensure that an acceptable risk exposure is obtained. One must also control whether the action in turn creates new risks. There are a number of strategies (see Section 6.11.3 and Figure 6.53).

6.11.10.1 Strategies for Threats

Avoid. Can the risk event be isolated? Can it be avoided by not implementing that part of the project or contract? For example, at the contract

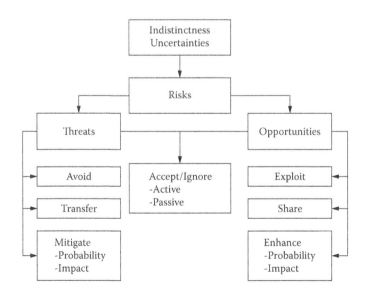

FIGURE 6.53
Risk response to threats and opportunities.

signing, the risk activity can be written off or clarified and formulated in such a way that the risk disappears or it can be made clear that negotiations with the authorities for environmental permits are not included in the project.

Transfer. If the risk is transferred to someone else, the risk remains but the responsibility is delegated. With banks and insurance companies, it is possible to insure against the threat of higher interest rates, currency fluctuations, and bankruptcy. If it is uncertain whether the resources and knowledge to perform a particular job exist, a subcontractor can be hired and guarantees used to "assure" that the performance criteria of the contract are met or at least compensation is given if the risk occurs.

The type of remuneration is a common way to transfer the economic uncertainty (see Section 2.6 in Chapter 2). It always costs to transfer the risk to someone else and this is not always the best solution. Contractors and suppliers want to get paid to take over a risk—sometimes a lot of money. Perhaps it is better to identify the risk and agree on how it should be regulated if it occurs. Too many extra reserves can result in the project closing down.

Mitigate. The probability or impact or both of these parameters are reduced so that the risk exposure comes down to an acceptable level. For example, one way to mitigate the risk of power failure because a cable accidentally has been dug off is to supply power from two directions.

One way to study the risk is to use the "what if" method. Remember the addition rule:

$$P(A \text{ or } B) = P(A) + P(B)$$

$$P(A) = A/(A + B)$$

Figure 6.54 shows the risk for fire and explosion in walk-in fume cupboards and a bunker where organic solvents may leak out. There is also a pressure tank with 20 million liters of N_2. It has been determined that a pressure-reducing valve or a safety valve may be wrong every 10 years. This yields the following scenario:

If there is *one* pressure-reducing valve (which experience indicates fails once every 10 years), then it may blow out 20 million liters of N_2 once in 10 years. The safety valve (which experience indicates fails once every 15 years) works the N_2 gas flows out into the air. If the safety valve does not work but the burst disc works, the N_2 and solvent will blow out into the

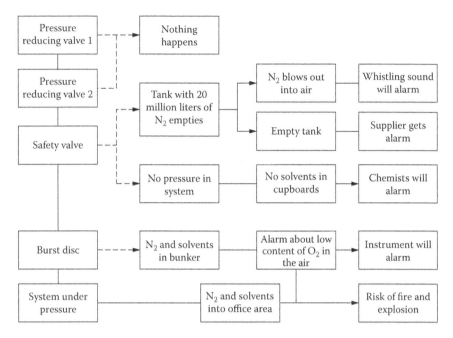

FIGURE 6.54
What-if example.

bunker. This will happen if both the pressure-reducing valve and safety valve fail the same month. The probability of this is once every 1,800 years—$(1/120) \times (1/180) = 1/21,600$, which is once in 21,600 months (1,800 years). If the blast disc does not work, the gas and solvent blow into shafts and the office area.

In this example, it is not acceptable that 20 million liters of N_2 can flow out once every 10 years. Therefore, a second pressure reducing valve is set up as shown in the figure. The probability of emptying the tank will then be once in 1,200 years $(1/120) \times (1/120) = 14,400$ months and that the gas and solvent blow into the bunker once in 216,000 years $(1/120) \times (1/120) \times (1/180) = 2,592,000$ months. Management will accept this solution, so it is chosen.

Accept—actively or passively. All risks cannot be eliminated from a project. Even if a risk is reduced or transferred, an acceptable part is left. One must accept that certain events may occur. Passive acceptance means that nothing is done and the risk is taken if and when it arrives. In active acceptance, something is done. If the impact has the highest value, a crash plan is prepared even if the probability is very low. The most common activity is to include reserves in the project. The reserves may be time, money, or human resources. Contingency reserves are for identified threats and management reserves for unidentified threats. Contingency reserves are expressed in hours, dollars, or people, while the management reserves are normally expressed as a percentage. Contingency reserves can be calculated using the expected monetary value method.

For some events, such as environmental and safety risks, crash plans are required. Although everything has been done to avoid an event, it can still occur. In the Öresund project, every safety helmet had instructions on what to do, where to call, etc., if an accident happened.

6.11.10.2 Opportunity Strategies

Exploit the opportunity. This may shorten the time or enable the work to be done more cheaply. For example, installing a temporary second elevator for transport of material can reduce traffic on already finished landings. Maybe material can be bought together with other projects and thus better discounts be obtained.

Share the opportunity with the client or third parties. Engaging a third partner and buying a special machine means that the job can probably be done more cheaply. Investment costs and savings are shared with the partner.

Enhance increases the likelihood that the opportunity will occur and/or improves impact, thus improving results.

Ignore (accept) the possibility. If nothing is done about the opportunity, it may still occur. It then becomes a pleasant surprise.

It is not enough to determine what action should be done and get the risk response money and reserves into the budget. The RR must be designated to someone who will be accountable for carrying out the action. All information should be entered into the RMP. Participants in the entire project must be informed of the measures that have been decided. Stakeholders should also be informed.

6.11.11 Risk Management during the Implementation Phase

Risk response has been planned and reserves have been based on uncertainties and risks. During the implementation, it is necessary to monitor the RMP continuously and to

- Perform risk response actions

- Monitor that actions are implemented
- Act if the actions have not been implemented
- Discuss and assess triggers
- Reevaluate probabilities and impacts, if conditions have changed
- Keep stakeholders informed
- Evaluate the effectiveness of the risk response
- Document experiences

Monitoring takes place at regular meetings. One person can be responsible for monitoring progress and report on meetings. Today's computer tools make it easy to organize actions chronologically for persons responsible for RR actions from the RMP. One concentrates on what can happen in the near future. Backward in time, the only discussion concerns whether the event happened and the RR activity worked. The risk/uncertainty point should not be the last one on the agenda because chances are that it will not be discussed at all. The progress report should include a paragraph that briefly describes what lies ahead in terms of risks and RR actions.

Triggers. With the help of project management tools, there are signals if there is a risk that the project will not reach its targets. Deviations in time schedules, budget prognoses, and many quality incident reports are clear signals that something must be done. Notice of strike, increased oil prices, currency fluctuations, new owners, and invoices that are paid later than agreed can be signs that a reassessment of probabilities and consequences is necessary. An increasing number of change requests signals that the quality of the project prerequisites is not good enough.

If relationships within the project or with the client, designers, contractors, or suppliers deteriorate, this signals a need for analysis and action. In all these cases, one must act to get control of the project. A professional PM not only notices what is happening but also acts to achieve the project objectives.

6.11.12 Evaluation and Documentation

Documentation is a communication tool that provides valuable reference data for future projects. It is an ongoing process that must be honest and concise. In the evaluation, whether the strategy had the intended effect or not is assessed. With that information, the RMP for the next project can be improved.

6.12 PROCUREMENT: PURCHASING AND DELIVERY CONTROL

WHAT should be bought?

WHAT type of contract should be chosen?

WHAT type of remuneration should be chosen?

WHAT kind of collaboration should be chosen?

HOW should buying proceed?

HOW will it be known if prospective suppliers can deliver the right quality and on time?

HOW will the price be determined?

HOW and **WHEN** should the seller get paid?

HOW can the architect's, engineers', contractors', and suppliers' meeting the contract requirements be gradually controlled?

HOW can the administrative requirements of the contract be met?

6.12.1 General

What are one's needs? Can one fix things oneself? Can anyone deliver things faster, cheaper, with less risk, and with sufficient quality than oneself?

In the construction industry, services (consultants/management), production (contractors), and equipment/supplies (suppliers) are bought. In all cases, there are a buyer and a seller. According to PMBOK®, the purchasing process includes three steps:

- Decide whether to do the work or purchase it. Plan purchases and acquisitions. Plan contracting and select sellers. Develop and send out inquiry specifications and inquiry/bid documents (preaward phase).
- Evaluate offer or bid, negotiate, and sign the contract (award phase).
- Administer the contract, inspect delivery and installation, commission, test on completion, create taking-over and performance certificates, and close contract (postaward phase).

All contracts are based on the fact that there is a need that someone else can fulfill. Sometimes sellers are trying to create a need that is not always a true one. Remember that a project or part of a project that is based on wishes rather than on needs is often hopeless. Most companies have a code of ethics that should be followed in the procurement work.

6.12.2 Plan Procurement

The buyer must ask:

- Should we do it ourselves? Do we have the knowledge? Do we have the capacity?
- Are there skills and resources that we can buy?
- Do we want to transfer the risk of failing to meet sub-budgets, deadlines, and quality standards to someone else?

The seller must first ask the following questions:

- Can we deliver what is requested?
- Do we want to deliver what is requested?
- Do we want to do business with the buyer?
- Do we take the risk?
- Do we want to establish ourselves in this region or this country?
- Do we go into this segment of business?

When considering purchasing, the type of delivery/contract, remuneration terms, and cooperation that are desired must be determined (see Sections 2.4–2.8). A simple way to plan the procurement is to do a procurement/purchase breakdown structure (PBS; compare WBS) and ask the questions what, how, when, and who. Each level can be broken down to more detailed levels (see Figures 6.55 and 6.56).

6.12.2.1 Purchase Package Content

The *delivery description* specifies what should be purchased, including amount, sizes, technical and functional specifications, and performance. Here the necessary **inquiry/bid documents** (drawings, technical descriptions, etc.) are specified. It is also necessary to specify the *assumptions* made when the purchase was planned and perhaps an explanation of why one is buying rather than doing the job oneself. One may want to explain why a particular form of *collaboration, type of contract,* or *limited number of bidders* was chosen. Which type of *procurement procedure* should

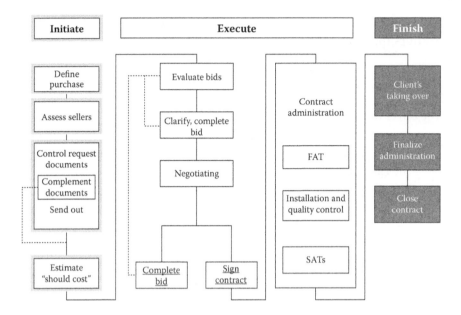

FIGURE 6.55
Procurement process.

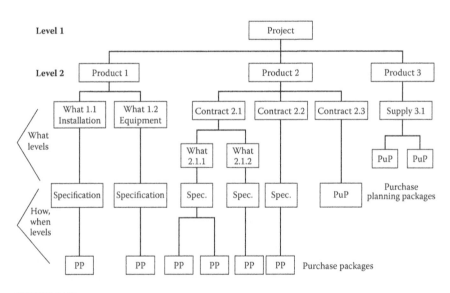

FIGURE 6.56
Procurement/purchase breakdown structure. Compare to WBS.

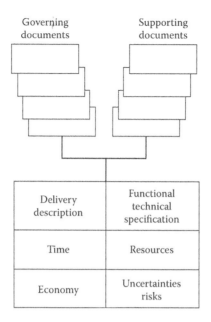

Governing documents   Supporting documents

Delivery description	Functional technical specification
Time	Resources
Economy	Uncertainties risks

FIGURE 6.57
Purchase package content. Compare to WBS.

be used? Note whether there are *strategies* to consider. This is important knowledge for the persons making the purchases. In terms of experience and know-how, this documentation gives something that may be needed in the future (see Figure 6.57).

Times? Which products (e.g., elevators, precast structures, and windows) have long delivery times and must be purchased early—perhaps separately? Note the latest day for contract signing that will give the contractors and suppliers a reasonable time for production.

What are the technical, legal, and financial *resources* needed to develop specifications and inquiry/bid documents and to evaluate bids?

What does the *money situation* look like before the purchase? What is in the budget? Which type of remuneration is preferable?

What are the *uncertainties, threats,* and *opportunities* associated with the procurement?

6.12.2.2 Purchasing Strategy and Tactics

When purchasing, it is necessary to know the project strategy and choose suitable tactics (see Section 3.5.2). What does the market situation look like? What is the contract's impact on the result? Are there many suppliers? Who has the knowledge and/or capacity?

6.12.2.3 Public Procurement

US authorities impose specific requirements on state and federal agencies, counties, etc., when they are buying goods and services. Similar requirements exist in Europe.

6.12.3 Delivery, Type of Contract

How can one find sellers with the right ideas and qualifications? In the procurement process, different contracting methods are used, depending on what is to be delivered:

- Request for information (RFI)
- Request for proposal (RFP)
- Request for quotation (RFQ)
- Competitive bidding
- One-step bidding (sealed)
- Sealed two-step bidding (technically acceptable solution at the lowest price when there is no clear specification)
- Noncompetitive negotiations
- Reverse auction

When **building contracts** are signed, sometimes both design and construction are bought (e.g., turnkey contracts). This depends on the type of contract (delivery form) chosen (see Section 2.5.2). This must be decided during the feasibility study.

When **design services** are bought, often not only design, drawings, and specification documents are bought, but also the scope management service. What are the needs of the users, project characteristics, and constraints? Sometimes the consultants help the client with scheduling, regulatory and environmental requirements, etc. (see Section 4.1.3).

In the first (design) phase of the project, functional, aesthetic, environmental, technical, and economic requirements and constraints (flexibility, operating costs, and technical sustainability) are determined. The second part of the design phase is to calculate the dimensions, make drawings, determine tolerances, and describe the test and monitoring functions. For a design–bid–construct (DBC) contract, the drawings and specifications are prepared by architects and engineers hired by the client and the contractors bid based on these documents.

6.12.3.1 Selection of Type of Delivery/Building Contract

Which parameters affect the type of construction contract? Who owns the knowledge and resources? What risks is one willing to take? What is the interest in bringing the development forward and trying new solutions?

Production knowledge is important in some projects where production methods greatly affect the price. An example of this is the tunnel under the Öresund (between Sweden and Denmark), where the bidders had different production methods and the offered prices differed greatly. In other projects, it is the technical solutions and the needs of the users that are most important (not the production method). When can the users specify their "frozen" requirements? For a turnkey project with a fixed price, the contractor is often more production oriented than looking for the best solution that benefits the project–user perspective. If the selection of subcontractors is important for the customer, CGC or divided contracts (see Section 2.5.2) might be chosen to ensure that the desired subcontractors are obtained.

Time and risk transfer parameters are important when choosing the type of delivery/building contract and remuneration.

6.12.3.2 Purchase of Products

Purchase of "products" means purchase of standard machines and products that are customized—for example, copying machines, kitchen appliances, tools, and vehicles. For many products, such as office supplies, the company has agreements and it is necessary only to specify and order the product. Prices and conditions are regulated in a framework agreement.

6.12.3.3 Purchase of Building/Construction

The following types of contracts exist in the construction industry:

- Concept contracts
- Functional/performance contracts (FuP)
- Design and construct contracts (DC)

- Design–bid–construct contracts (DBC)—also called engineering, procurement, and construct (EPC) contracts
- Coordinated general construction or divided contracts
- Management organization contracts
- Subcontracts (SCs)
- Outside or joint contracts (project work not included in the contractor's scope)
- Supplier contracts

More information about the different types of contracts can be found in Chapter 2, Sections 2.5.1 and 2.5.2.

6.12.3.4 Contracts with Architects and Design Engineers

Before the decision to hire architects and design engineers is made, it is necessary to consider **when** the project can be influenced most. The possibility to influence the project is greatest at the beginning. As the project develops, there are fewer possibilities. The work intensity (drawings and building) will increase with time, as shown in Figure 6.58. The black curve shows the economic turnover of the project, which of course is greatest during the last phase of production, when all subcontractors are involved. The dashed curve shows how much the project can be influenced. Of

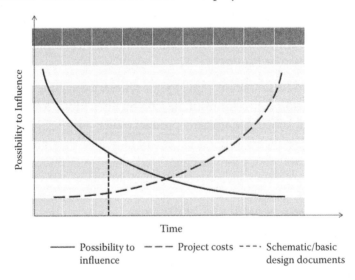

FIGURE 6.58
The ability to influence the project cost (black line) versus costs (dashed line).

course, all the really smart thinking should take place in the beginning. The dashed vertical line represents the schematic/basic design documents.

Because the ability to affect a project and find the smartest solutions is at the beginning, the most creative, knowledgeable, and experienced consultants should be used then. To lock up the consultants with fixed prices in this early stage is not very creative. Instead, the client and the consultant should discuss a budget for the feasibility study and the system/basic design documents. The consultant should also provide a budget for the building documents based on the knowledge available in the very early stages. There should also be a contract text that allows the client to use the feasibility study and system/basic design documents with other consultants if an agreement for the price of the inquiry/building documents cannot be reached or a type of contract is chosen in which the contractor is responsible for the design and chooses its own favorite consultants.

Bids from design engineers should be split into different phases. The agreement can have two phases: feasibility study and system/basic design documents (step 1) and inquiry/building documents (step 2). For the first phase, the consultant will submit a monthly budget for the feasibility study and system/basic design documents because, after these documents are submitted, the building documents that are needed are known and a fixed price can be used. For the inquiry/building documents, a fixed price based on the bid for step 2 can be used, with adjustments for changes and better knowledge of the project. Do not forget to include in the contract a clarification that the client can hire a new consultant after the first phase if agreement on the compensation for the construction documents cannot be reached. The fixed price should include compensation for changes caused by missed coordination with other consultants. Any additional costs due to changes that depend on other consultants should be regulated directly between the consultants.

Choosing a consultant is difficult. If one offer shows 400 hours at $250 per hour and another 500 hours at $200 per hour, which should be chosen? A discussion and breakdown of the hours in smaller parts can help in making the decision.

6.12.4 Request/Bid Documents

The statement of work (SOW) defines the portion of the project scope that should be included in the contract. When the inquiry/bid documents are prepared, standard terms and conditions should be used. The type of contract desired should be chosen and model agreement forms used. A number jointly prepared by sale and purchase organizations (e.g., FIDC) can be used (see Section 2.5.1).

Bid documents should be based on general and particular conditions and standard and particular specifications. Standard specifications have been prepared from technical requirements in existing standards of major industrial users and contractors. *Standards* are considered voluntary because they serve as guidelines and do not have the force of law. A *code* is a standard that has been adopted by one or more governmental bodies and has the force of law. "Building standards" details the type, quality, and color selection of design materials and finishes. Examples of building standards include tile, carpet, paint and wall coverings, baseboards, counters, plumbing and lighting fixtures, and other materials needed to complete the design aspects of a finished space. The American Society of Mechanical Engineers (ASME), for example, accredits users of standards to ensure that they are capable of manufacturing products that meet the ASME standards. It provides stamps that accredited manufacturers place on their products, indicating that a product was manufactured according to a standard.

ASME cannot, however, force any manufacturer, inspector, or installer to follow ASME standards if this is not written into the contract. The American National Standards Institute has compiled a list of standards that improve productivity, increase efficiency, and reduce cost. Other standards are British standards, European standards, and international standards (ISO). Process industry practices (PIPs) have standard specifications (e.g., for piping material).

6.12.5 Buyer/Supplier Assessment

Before a business deal is entered into, both the buyer and the seller must know or find out the following about the counterpart:

- Overall business strategy
- Technical knowledge
- Capacity to deliver the right quality in time
- What are their key objectives?
- Reputation
- Organization structures for projects
- The financial situation of the counterpart

- What are their difficult contract parts for the opposite party? Can we help them?

The financial situation of the counterpart, before sending out documents, the client's purpose of supplier assessment is to clarify the following:

- The bidder has the financial, technical, and human resources to carry out an assignment on time according to the technical requirements of the contract.
- The bidder has the time and willingness to bid for the current project.
- There is the requisite number of bidders for healthy competitive bidding.
- Bidders who cannot be accepted as suppliers do not spend time and money on unnecessary bid work.
- Rogue bidders do not complicate the evaluation process.
- Bid evaluation can be fast.

6.12.5.1 Activity Description of Supply Assessment

Assess and supplement the existing list of prospective bidders to provide alternatives and competition by

- Checking for a general agreement between the company and its bidders
- Determining who may be accepted as a contractor
- Finding new, supplementary bidders

Obtain and evaluate current information on

- Financial position
- Experience from similar assignments
- Prospective material resources, capabilities, and competence of personnel (to be completed during the negotiation phase)
- Adherence to schedule and budget in previous projects
- Business conduct in connection with changes and additions to work in previous projects
- Quality and environmental management

If possible, visit and/or conduct quality audits and carry out risk analyses of the current consultants, contractors, and suppliers.

6.12.5.2 Comments and Recommendations for Supply Assessment

Sometimes, to create price competition, one can pretend that there are more candidates than there actually are. However, one should not lie about this.

6.12.6 Evaluation of Bids

When offers have been collected, one bidder should be selected. Although price may be the primary determinant, technical merit, past performances, time and payment schedules, SHE policies, and management must also be considered. The offered price should be weighed against an independently estimated "should cost." Expert judgment on technical solutions and legal assistance may be needed. If there are outstanding issues, additional information may be required. Bids are normally only valid until a certain date and sometimes bid evaluation is late. One type of bid completion may be to ask the bidder to extend the bid time. Watch for how long the offer is valid. To evaluate bids, selection criteria must be identified and specified. Top management and stakeholders must understand these criteria. If they do not accept them, they have to react.

Clarify any ambiguities in the bid documents. Ambiguities must not be present at the final hearing or contract writing. Compare bids with the budget and comment on possible causes of deviations from budget and "should cost." Assess the risks associated with the bid. Remember that risk analysis should not only see the threats, but also define the opportunities.

When bids are opened, at least two people should be present. (Of course, this is not valid for one's own projects.) Vital information in the different offers should be documented in a "bid-opening document." This document should be signed by at least two persons. Bids and the bid-opening documents should be stored in a safe, fireproof location.

Many bids can contain wording that will describe the risks that may occur to the other party. These are often included to limit the bidder's risk. The goal should be to discuss the risks and agree on who should take the risk and how it should be regulated if it occurs.

6.12.7 Negotiation

A project is limited not only by time, cost, and scope but also by relationships between client and project members. Can one afford to damage the

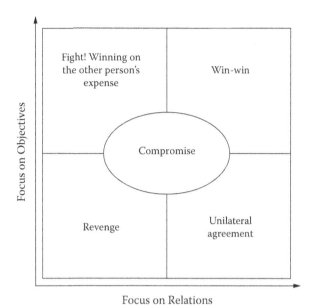

FIGURE 6.59
Negotiating strategies.

relationship with a counterpart? The answer is no. Negative information is disseminated quickly. Consider what the proverb says:

> *"The rumor travels faster between Baghdad and Damascus than it takes for you to put on your sandals."*

Negotiation strategy is described in Figure 6.59.

The golden rule of any negotiation is never to lie! It is not necessary always to tell everything that is known, but never lie. Think of reputation. Who knows? In 2 years, one's counterpart may be one's boss. Does one want him or her to remember a liar or a skilled negotiator? Agreements are primarily based on trust and, secondly, on a well-written contract. The buyer and seller have their own objectives and interests. Some of these objectives are conflicting and some are not. Be especially vigilant concerning requirements and offers that are not normal in the industry. Clarify the motives and risks.

The purpose of negotiation is to ensure the following:

- The contract is unambiguous and cannot give rise to interpretive disputes.

- A contractor who can fulfill the contract to the best possible conditions is obtained.
- A schedule that can be checked against is accepted by both parties.
- Financial tools are in place to regulate the changes and additions.

The following negotiation activities should be included:

- Create a good climate for negotiations with a possible win–win situation in the introduction.
- Clarify what is agreed on.
- Define what is not agreed on.
- Compile in writing what has been agreed on.
- Discuss outstanding questions.
- Agree on a continued negotiation process, if there are remaining issues.
- Document.
- Sign contract.

The following points should be addressed and clarified at the meeting:

- Outstanding issues from previous negotiations and bid additions
- Review of all the points in the contract form

Many times it is appropriate to prepare a preliminary contract, which can then be negotiated. By doing this, the preparer takes the initiative in the negotiation and makes sure that his or her issues will be discussed.

Negotiation is the art of the possible. Think win–win. An issue can be avoided, but never lied about. It is not necessary to tell everything. The other party has an obligation to find the items that could be administered to his or her advantage.

Technical and legal matters of detail should not be addressed in the discussion unless one is a specialist and has studied the bid documents carefully at this point. Get help if anything is not understood. Ask to come back for another discussion.

Negotiations are basically the effort to maximize the chances that one party will decide what *the other party* wants it to decide. And one party should let the other party get what the first party wants it to get. One should put oneself in the other party's role. As a buyer, be aware that the opponent also has a need, wanting to sell its goods, employ its labor, or enter a new market. It is win–win that applies in today's negotiations.

A good negotiator can effectively reach or come close to realistic negotiation objectives. It is not a good negotiation if one party leaves the table happy, while the other party feels steamrolled. The other party must feel that it got something. If not, the relationship is in danger.

The following thoughts are not unusual: "One day, I'll tie up your tail for this" or "if I only get the contract, I will ensure that the price will be right. There will be more negotiations, over changes, when the buyer's situation isn't so strong."

The key words in today's negotiations are win–win. This does not mean that everything will be divided 50/50. It may well be 95/5 to one party, as long as the counterparty has had one or more of its important needs met. How are the other party's needs found? By questioning, listening, and analyzing. Try to find simplifications that also benefit the counterpart. Present the offer as a positive one.

Sometimes it is possible to prepare by taking the counterparty's role. This may uncover weaknesses in the opponent's reasoning. Try to understand the other party and its needs. What is to be done if the negotiations fail?

6.12.7.1 BATNA

A common mistake is to say no to an agreement that is better than the BATNA (best alternative to a negotiated agreement) or to say yes to something that is worse than the BATNA. What is the counterparty's BATNA? Note that the BATNA is *not* someone's negotiating objectives. Unfortunately, it is documented that negotiators who know the BATNA do less well than those who only know their negotiating goals. A satisfied feeling arises when a result that is better than BATNA is achieved. The negotiator must therefore try to "forget" BATNA and focus on the negotiating objective.

Contract negotiations are often complex. Time, scope/performance, costs, information sharing, risk sharing, outside contractors' work, etc., are issues that need to be clarified. The discussions should not be "arm-wrestling," where one party is the champ and the other a loser. Therein lies the negotiator's dilemma: how much information to give the other party without being exploited.

Best negotiating results are usually achieved if both parties have good knowledge of negotiation skills, win–win, objectives, priorities, and preferences and understand the negotiation process and the possibilities, instead of thinking in terms of prestige.

Confucius said,

"Anyone who knows many is strong, but he who knows himself is stronger."

Make sure to be regarded as an equal and skillful negotiator.

6.12.7.2 Prepare to Negotiate

Some basic things must be done before negotiating: prepare, prepare, prepare... Plan the negotiations and try to take the initiative:

- List deliverables and their prices (budget, offer, worst case, most likely, best case).
- What are the milestones? Which ones can be changed and by how much?
- Which positions in the counterpart's organization are so important that a certain person should be named and included in the contract?
- What subcontractors should be known about and maybe included in the contract?
- How will the inspections, final inspections, and taking-over certificate be administerated?

- What is wanted or can be accepted regarding the payment schedule?
- Are there any approvals before the next step in the production can take place?
- What are the negotiator's personal objectives?
- Are there other important issues?

- Know the technical SOW and the not-technical SOW.
- What is the preferred remuneration? Can any other type be accepted?
- What are the difficult contract parts for the counterparty to accept and deliver? Can they be helped then?

- Are there any ambiguities that need to be clarified? What is my opinion on these issues?
- What permits, licenses, or approvals must be obtained? Which are needed? Who is responsible for the contacts with different authorities?
- Bonds, guarantees, and insurance need to be addressed.
- Who is the negotiator? Does he or she have the authority to negotiate?
- How is the SHE work organized?

Prepare a negotiation agenda before starting. Think of the preceding points as well as the following ones:

- Introduction of negotiating team members (both parties)
- SOW/control of contract documents that will be part of the contract; priority of documents
- Clarification of any remaining ambiguities
- Contract terms and conditions
- Experience and capacity; past performances
- Milestones and schedule
- Confirmation of right of access to the site and access routes
- Client's and seller's delegation of powers
- Subcontractors
- Who is responsible for electricity, water, and gas on-site
- Change order process
- Force majeure
- Claims, disputes, and arbitration

- Clarification of the type of contract or part of a contract that will be discussed
- SOW/functions and quantity

- SOW/administration, services, and technical issues
- Liquidated damages
- If services and/or production, organization structure; specially named personnel
- Payment schedule; advance payments
- Confidential details; intellectual rights

- Bonds, securities, and guarantees
- SHE organization and work
- Security on-site

- Defects liability, guarantees
- Insurance
- Taking-over procedures

Sometimes one will be confronted with win–lose negotiating tactics. The best thing is to learn about such tactics (e.g., insulting behavior, citing false data, arbitrary deadlines, denying authority, stonewalling) and use the recommended countertactics.

6.12.8 Securities, Bonds, and Insurance

6.12.8.1 Securities and Bonds

In construction projects there are always risks. One primary risk is that contractors and/or suppliers will prove unable to complete a project or will fail to pay all subcontractors or suppliers. Another primary risk is that the buyer cannot pay contractors or suppliers. Project owners need assurance that the bidder they select for a job will be able to complete the job properly.

With bonds, the owner has assurance that the bidder has the financial means to accept the job for the price quoted in the bid. The use of surety bonds in construction projects acts to shift and spread certain risks that owners on construction projects face. The risks are partly transferred to banks, insurance companies, etc. Some common securities and bonds are (1) bid bond, (2) performance security/bond, and (3) payment bond.

During the bid evaluation phase, a bid bond helps the owner to assess if a contractor has the financial means to perform the job. A bond is a financial document that must be handled in a proper and secure way. Be sure to include in the contract that the first invoice will not be paid until contracted bonds are in the hands of the client. Some securities and bonds will not be returned until the guarantee period is over.

6.12.8.2 Insurance

- **Purpose**
  - Ensure that the contractor has financial protection and does not go bankrupt in order to remedy the damage caused by, for example, a fire.
  - Ensure that the contractor has financial protection and does not go bankrupt and can fix the damage on existing buildings where the renovation, conversion, or extension is performed.
  - Ensure that the contractor has financial protection from third-party claims that can be brought against it.

Include in the contract: "The first invoice will not be paid until proof of insurance is obtained and given to the buyer."

The documents should be kept safe. Remember to check that the contractor has paid the insurance premium. In a much-publicized project, a foreign bank stated that it had not received the premium for the last 2 years and claimed, to the client's horror, that the presented insurance did not apply.

Work with insurance includes

- Checking that the policy meets the contract requirements
- Reviewing and recording the validity of length of the insurance period
- Monitoring validity of the insurance and ensuring that new evidence is received that covers the entire contract and warranty period

Keep track of all securities, bonds, and insurance documents. Have a log that indicates where they are, how long they are valid, and to whom they should be returned.

6.12.9 Contract Administration in the Design Phase

- **Purpose**
  - Ensure that consultants' documents are complete and coordinated with outside consultants.
  - In TK (turnkey) and DC (design–construct) contracts, ensure compliance with contract requirements.

The consultants' performance in relation to the contract involves the following:

- Review that the consultants' discretionary inspection is carried out and spot-check that the consultants' control plans and checklists for the project are used.
- Review request documents before they are sent out.
- Conduct quality audits.
- Control that the functional requirements are not distorted during the design phase.
- Review and critically follow the schedule and act on noncompliance.
- Manage contract changes through agreed processes.
- Implement the risk management process for the design work and proposed technical solutions.
- Review the SHE requirements.
- Inform each other of the design development of specific technical solutions.

6.12.10 Contract Administration of the Production Phase

- **Purpose**
  - Ensure that the contractor's/supplier's product at the final inspection is complete and in accordance with the contract and coordinated with outside contracts.
  - Conduct normative inspections to ensure that customer and supplier interpretation of the standard requirements is consistent.
  - Early on, ensure that the client's *intended* standards are met. If the contract says something else, this may mean additional orders.

6.12.10.1 Construction and Installation Control

At a construction site, there are several players, such as general contractors, subcontractors, outside contractors, and suppliers. All players supply material to the workplace where products are assembled. To reduce the risk of production disruption, the goods that arrive for the assembly must not be faulty or damaged. They must be stored and installed in a way so that they are not damaged.

Drawings are produced by many different consultants and it is not uncommon for there to be a lack of coordination between drawings. This creates problems during the production stage. In a TK/DB contract, the main contractor is responsible for coordinating the drawings; however, for DBC and GC contracts, the client is responsible for the coordination of drawing. In order to avoid additional costs of production, it is necessary to act quickly and resolve any drawing coordination problems.

Sellers and buyers must agree on what is in the agreement. If the agreement is unclear and the seller intends to deliver one product, but the buyer is of another opinion, this problem has to be solved very quickly. If there are different interpretations of the contract text, calling the inspectors of completion, who can resolve this by a normative inspection, is recommended.

Construction work contains many interfaces between different contractors and installers. Work must be done in a certain order. The production system is controlled by the main contractor or the general contractor. With a divided contract, the client is responsible for the production coordination. This requires skills, experience, and planning knowledge. Interference may result in financial implications (e.g., unplanned delays). Always try to transfer the responsibility for coordinating to the main contractor.

When similar rooms (e.g., office buildings or hotels) are repeated and the installation space is tight (e.g., hospitals, research laboratories), building a prototype room is recommended. This will help the client to understand what has been bought and the contractors to be aware of coordination matters, working methods, and assembly of fittings and furnishings.

6.12.10.2 Commissioning of the Plant

Commissioning of new facilities is often done gradually during the time for completion (and attributed time to the contractor). The operating and maintenance departments should be informed well in advance as this may affect an existing operating facility. In the final stages of the contract work, heating from the new system is frequently used. This is done to avoid polluting and bulky temporary heating devices inside the new premises. Permission to commission the permanent heating system must be entered into the contract.

Commissioning of the ventilation fans should not take place until a SAT 1 (see Section 6.13.1) has been performed for all subcontractors and outside contractors. The equipment that they install can be contaminated by construction debris from the fan rooms and channels. New filters should be installed after the coordination of the ventilation system has been tested and before the final inspection. Experience shows that the filters often are clogged with construction dust.

Installation of machinery, equipment and furnishings must be done in close cooperation between client, contractors, and suppliers. It is often useful to test the product at the factory (FATs) before it is transported to the site. Often it is both cheaper and faster to get something fixed at the factory than on the site. Notification of delivery of machinery, equipment, or furnishing must be occur in sufficient time so that the site organization can plan for the reception. Unloading, transportation, storage, protection of surfaces, and transfer of responsibility for the product should be regulated in the contract. Reception inspection must be planned and specified.

6.12.11 Normative Inspections and FATs

6.12.11.1 Preliminary Inspections, Normative Inspections, and Prototype Installations

- **Purpose**
  - Inspect the site before construction work starts.
  - Ensure early clarification of contractual legality of specific design.
  - Ensure that customer and supplier interpretation of the standard requirements is consistent.
  - Avoid faults and defects remaining at the final inspection.
  - Ensure that contractors receive early knowledge of installation complexity and learn about the time needed for certain work where many contractors have jobs to do.
  - Ensure that parts that are hidden during further production are inspected and approved before they are built in.
  - Give users an opportunity to respond to the design and execution (e.g., surface finishes).

Description of activities. Before construction work starts on a site, a documented inspection of the work area should be done and any damage to adjacent buildings noted. Sometimes this requires a more detailed inspection of the adjacent buildings. This is done, for example, before blasting operations or when settlement and damage to nearby buildings is possible.

The protocol should note any deviations from the contract with regard to the conditions for the contractor at the start of production. Minutes should be approved at the next construction meeting.

Preliminary inspections should be done for building components, installations, fittings, and seals (with or without fire requirements)

- Prior to closing the enclosed shaft
- Prior to closing the fixed ceiling (e.g., gypsum board)
- Before energizing the switchgear
- Prior to taking over parts of work or outside contractors' installations of machinery, equipment, and fittings

Inspections are performed for early detection of faulty or ambiguous performance—for instance:

- Building work
  - Measurement of floor tolerances and moisture content before flooring; coating, layer, floor covering
  - Performance of the first built part of the roof and facade; inspecting to make sure they are leakproof and sealed
  - After the first holing through walls and floors, checking seals and refinishing
  - When work with special requirement starts
- Mechanical work
  - After the first pipes are installed
  - After the first insulation works are installed
  - After the first fixing of studs with requirements is installed
  - When the first gas pipes are welded
- HVAC work
  - After the first ventilation ducts are installed
  - After the first insulation works are installed
  - After the first fixing studs with requirements are installed
  - When the air handling units are installed

- Electrical work
  - When the first conduits and cable racks are installed
  - When the first cable installations are done
  - When the first electrical distribution board is installed
  - When the first light fittings are installed

Sometimes, during production, disagreement between the contractor and the client may occur. This problem should not be postponed to the final inspection. Rather, the dispute should be resolved when it occurs.

6.12.11.2 Factory Acceptance Test

- **Purpose**
  - Control that the contract requirements are met before deliveries are shipped from the factory.
  - Control that deviations are taken care of with the right equipment and the right physical conditions.
  - Control that the facility is tested with the existing equipment at the factory (if it is impossible or expensive to move it to the site).

Products shipped from the factory should be flawless and complete. Usually it is easier to correct a fault in the factory than outdoors on a cold and rainy day at work. Steel beam welds, gusset plates and stiffeners, thickness of galvanic protection, etc., should be checked before delivery and the certificates should preferably be sent with the delivery. Electrical substations, switchgear, and transformation stations should be tested before delivery to the site.

An error-free product that does not need to be replaced or rebuilt on-site reduces the risk of delays. Testing can be carried out as visual inspection, sampling, review of the supplier's control, and collection of documents. The equipment provider should perform factory acceptance tests under the supervision of the client. Requirements for FATs should be included in the contract.

6.12.11.3 Description of FAT Activity

- **Start-up**
  - Notice of FAT that describes the purpose of the tests, test sites, gathering, and how long the test is expected to take

- Planning of human resources and equipment needed to perform the test
- Schedule for testing
- Checklists for testing
- Physical conditions for testing
- Description of test method, step by step
- Opening meeting
- Checking that the reference materials are in place
- **Implementation**
  - Component control, control of quality and quantity (sometimes called FAT 1)
  - Control that the software and the internal systems of the product work (sometimes called FAT 2)
  - System test: verifying that the system works together with external systems through simulation of the external interface (sometimes called FAT 3)
  - Verification of operational and maintenance requirements (sometimes included in FAT 3)
  - Control of preliminary as-built documentation and all operation and maintenance manuals available to the client to be completed with information that becomes available later
- **Finish**
  - Compilation of materials
  - Closing meeting
  - Reporting

Should the customer, client, or project owner be called to FAT? Remember that the result is the product of quality and acceptance (R = Q × A). If customers have any comments on the product at this point, there is now a "last chance" for additional orders in the form of amendments or additions. If a representative of the client is present at the test, it can be a good idea for him or her to sign that the product, in his or her opinion, is approved.

6.12.12 Claims, Disputes, and Arbitration

If one signs a contract, one is bound by the agreement. Both the client and the designer, contractor, and supplier must fulfill their obligations. If any of the parties violates the agreement, the injured party may seek damages. There are two major categories of monetary damages:

Compensatory damages are meant to put the victim in the same monetary position as though the violation had not occurred. The purpose with *punitive damages* is to punish the breaching party, not to compensate. Contracts are normally limited to compensatory damages.

Liquidated damages allow for compensation for damage where the sums are agreed on in advance (e.g., $1,000 per week if the project is delayed). The terms are normally included in the contract. Late deliveries are often connected with liquidated damages.

Many times, a dispute adjudication board (DAB) is appointed jointly by the client and the contractor. If a dispute between the parties arises, it can be referred to the DAB, which then decides the issue. If any of the parties is dissatisfied, it can be possible to settle the dispute in court or under the Rules of Arbitration of the International Chamber of Commerce. Text about the procedures should be included in the contract (see FIDIC, Conditions of Contracts for Construction. § 20 Claims, Disputes and Arbitration).

6.13 APPROVAL MANAGEMENT

WHAT should be done to get the contract/project approved?
WHAT regulatory inspections are needed?
WHAT documentation and training should be delivered?
WHAT should be done administratively in connection with approval?
HOW is the process of approval and taking over carried out?
HOW can errors and deficiencies identified in inspections be corrected?
HOW are the administrative requirements of the contract approved?

There are two areas of focus at the end of the project: technical and administrative approvals. Technical approval concerns having a facility that can be used as intended (objective and functions). Administrative approval establishes that the financial and contractual terms (warranties, guarantees) end and that the contract and project are passed on to other responsible parties (users, operating department, etc.).

- **Purpose**
  - Ensure that the project is complete and without faults when it is delivered to client or users.
  - Ensure that the required regulatory inspections are done before the operations in the building start.
  - Ensure that the functional requirements and user requirement specifications are met.
  - Ensure that experience feedback from the project is documented.
  - Clarify who has the responsibility for the new facility when the operational department takes over.
  - Ensure that the warranty and how to exercise it are known.
  - Ensure that the automated systems are CE marked (European Union) before takeover.

Delivery control of the contractors and suppliers in the final stage is normally done through inspections in accordance with the contract or standard contract. In addition to these inspections, the following occur:

- Fire inspections
- OSHA inspections
- Final cleaning
- Replacement of lock cylinders
- Trimming of HVAC installation (e.g., heating and ventilation in winter and summer)
- Preparation and submission of as-built documents
- Delivery of operating and maintenance instructions
- Education (if included)
- Performance tests under normal production (depends on type of contract)

6.13.1 Construction and Installation Inspections, Performance Certificate

Final inspection and tests on completion must be done in the way described in the contract. A normal way is in the following steps (this procedure can be compacted into small projects):

- **Start-up**
  - Notice of SAT that describes the purpose of the tests, test sites, gathering, and how long the test is expected to take
  - Planning of human resources and equipment needed to perform the test
  - Schedule for testing
  - Checklists for testing
  - Physical conditions for testing and test equipment
  - Description of test method, step by step
  - Opening meeting
  - Checking that the reference materials are in place
- **Implementation**
  - Component control, control of quality and quantity (sometimes called SAT 1)
  - Control that the software and the internal systems of the product work (sometimes called SAT 2)
  - System test verifying that the system works together with external systems (sometimes called SAT 3)
  - Verification of operational and maintenance requirements (sometimes included in SAT 3)
  - Control of as-built documentation and that all operation and maintenance manuals are available to the client
- **Finish**
  - Compilation of materials
  - Closing meeting
  - Reporting

If the work passes the tests on completion, the client issues a taking-over certificate and then takes over the facility. If the work or parts of the work are found to be defective or otherwise not in accordance with the contract, the client may reject the facility. The contractor should then promptly correct the defect and ensure that the rejected items comply with the contract. When the defects are remedied and approved, the client issues a performance certificate, which constitutes acceptance of the work.

6.13.2 Administrative Finish of the Project

WHAT should be done before the project ends? **HOW** is the project finished correctly? Although the project will be technically approved, the following points must also be addressed:

- Money situation
- Regulatory requirements
- As-built documentation
- Feedback of experience
- Transition to the users
- Transition to the technical operation/maintenance team
- Show appreciation for the hard work and sacrifices made by the project members
- Final report

Make a project evaluation, assignment evaluation, etc.:

- **Purpose**
    - Ensure that **all** contract requirements are met.
    - Ensure that the outstanding liability and warranty issues with contractors or suppliers are finished.
    - Ensure that economic relations with contractors/suppliers are terminated.
    - Ensure that regulatory requirements are met.
    - Document the experiences:
        - Quality
        - Time
        - Costs
        - Technical solutions
        - Risks and uncertainties
        - Environmental and occupational experience
    - Maintain good relations between the project participants for future cooperation
    - Reduce or return securities or bonds
    - Document the project's final cost

One common problem is to get contractors and suppliers to quickly fix the defects noted at the tests on completion. Not paying the final invoice until all defects have been addressed and approved can be an excellent instrument to use to resolve the issues. Text about this must be included in the contract. It can be a good idea to have experience meetings with designers and contractors to learn about good and less than good technical solutions.

Party. Many people have made an effort to make the project a success. Conflicts have been resolved, sometimes with strong emotions. Overtime work has resulted in strained relationships with family. Now the work should be acknowledged and celebrated. Of course, the size of the party must be adapted to the size and difficulty.

Common Acronyms

A: Acceptance (A) in the formula R = A × Q

AC: Actual cost

BAC: Budgeted cost at completion

BIM: Building information management; in some countries, building information model

BoM: Bill of material

BoQ: Bill of quantities

BOOT: Build–own–operate–transfer

BOT: Build–operate–transfer

CGC: Coordinated general construction (contract)

CM: Contracts manager (see Section 2.1 in Chapter 2) or construction management (see Section 2.8)

CPI: Cost performance index

CPIF: Cost-plus-incentive fee

CPM: Critical path method

CPSC: Client-purchased side contracts

CRR: Comparative risk ranking

CV: Cost variance

DAB: Dispute adjudication board

DBC: Design–bid–construct contract

DC: Design–construct contract

DCM: Design–construct–maintain contract

DDC: Develop, design, construct contract; sometimes called guided design construct contract

DM: Designs manager (see Section 2.1)

ED: Enquiry document

EMC: Electromagnetic compatibility

EPA: Environmental Protection Agency

EPC: Engineering, procurement, construction contract

ESI™: ESI International; an Informa business with education in project management

EV: Earned value

F: Frequency

FBS: Functional breakdown structure

FDA: Food and Drug Administration

FIDIC: International Federation of Consulting Engineers (Fédération Internationale des Ingénieurs-Conseils)

FPIF: Fixed-price incentive, firm

FPO: Fire prevention officer

FuP: Functional/performance contract

GC: General construction (contract); general contractor

GDC: Guided design construct contract; sometimes called a develop, design, construct (DDC) contract

GLP: Good laboratory practice

GMP: Good manufacturing practice

HVAC: Heating, ventilation, and air conditioning

I: Impact

ICC: International Code Council

IM: Installation manager (see Section 2.1)

IRL: Interface responsible list

ISO: International Organization for Standardization

LR: Legal representative (agent)

MIS: Maintenance information system

NBC: National building code

NFPA: National Fire Protection Association

NGT: Nominal group technique

OSHA: Occupational Safety and Health Administration

P: Probability

PBS: Procurement/purchase breakdown structure

PC: Project charter

PERT: Program evaluation and review technique

PIPMO: Professional independent project management organization

PM: Project manager (see Section 2.1)

PMBOK®: Project management body of knowledge

PMD meeting: Project manager's decision meeting

PMI: Project Management Institute, Inc.

PPC: Public–private cooperation

PPP; P3: Public–private partnership

PQM: Project quality manager

PV: Planned value

Q: Quality (Q) in the formula $R = A \times Q$

QM: Quality manager

QS: Quality surveyor

R: Result (R) in the formula $R = A \times Q$

RBS: Risk breakdown structure

RFP: Request for proposal

RFQ: Request for quotation

RMP: Risk management plan

RPN: Risk priority number

R_x: Risk exposure

RR: Risk response

SBC: Standard building code

SBS: Scope breakdown structure

SC: Subcontractor

SHE: Safety, health, and environment

SM: Site manager (see Section 2.1); sometimes called team leader

SPI: Schedule performance index

SU: Supplier

SV: Scheduled variance

TK: Turnkey contract

TL: Task leader (see Section 2.1 in Chapter 2)

UBC: Uniform building code

URS: Users requirement specification (validation process)

Appendices

APPENDIX A:
PROJECT MANAGEMENT PLAN

CONTENTS

1 Plan Approval

This project management plan has been established for: New laboratories for organic chemistry

Drawn up by:
PM: Chris Strong Project

Approved by:
Owner: Anne Harding

1.0	General	
This project management plan follows the guidelines of Gadget Corp.		
1.0.1	Project name	New laboratories for organic chemistry. (part of building A)
1.0.2	Project owner	Anne Harding
1.0.3	Property owner	Gadget Corp.
1.0.4	Background	The developments of new drugs are promising and the research in organic chemistry needs better facilities.
1.0.5	General description	"Describe the project in a general and clear way"
1.0.6	Important stakeholders	
1.0.7	Strategies	
1.0.8	Environmental policy	See policy of Gadget Corp.
1.0.9	Others	People working with the chemistry processes must sign secrecy documents.
1.1	**Authorities, codes, permits**	
1.1.1	Description	Building permit document shall be prepared within 2 months.
1.2	**Decisions, integration**	
Meetings will be held in accordance with the project manual of the Gadget Corp.		
Special coordination meetings with chemist representatives will be held every third week.		
MF Chemical Consultants Corp. is responsible for specifications, purchasing, and installing of fume cupboards.		
1.3	**Scope**	
Work shall be done in accordance with the project manual of the Gadget Corp.		
Outline specification: See memo from PYZ Architects.		
Functional requirements: See memo from MF Chemical Consultants.		
Project limitations: See memo from Projsam ab.		
1.4	**Change management**	
See the project manual of the Gadget Corp.		

1.5 Environmental and work environmental management

Work shall be done in accordance with the project manual of the Gadget Corp.

Environmental manager for the project: Rose Bird

Client's work environmental manager: Robert Suspender

Project limitations: See memo from Projsam ab.

1.6 Quality management

Work shall be done in accordance with the project manual of the Gadget Corp.

MF Chemical Consultants Corp. is responsible for receiving Inspection for fume cupboards.

1.7 Time management

Project start	10-11-2011
Feasibility study ready for review	01-18-2012
System/basic design ready for review	02-20-2012
Application documents for building permit ready	03-20-2012
Frozen layout	04-20-2012
Enquiry documents ready	08-06-2012
Bids in	09-03-2012
Start construction	End of September 2012
Tests for completion	End of June 2014
Moving in	08-01-2014

1.8 Cost management

Investment proposal for detailed design when system/basic design is ready.

Investment proposal for production 07-02-2012

Countercalculated bids before opening sealed bids

Invoices shall be marked "Proj 123. New Lab OC" and Att. Paul Cash

1.9 Resource management

CAD manual approval	Beginning December 2011

Human resources

Organization structure	See the project manual of the Gadget Corp.

Client organization

Client representative (project owner)	Anne Harding
Project manager	Chris Strong
Users' representative	Evanca Smart
Security officer	Bill Fargo
Fire prevention officer	Carl Johnson
Environmental manager	Rose Bird
Work environmental manager	Robert Suspender

IT and communications manager	Bill Jacobs
Plant operation manager	Pat Tool

Design consultants (TL)

Architect Tobias Lynx	PYZ Architects
Structural engineer Thor Beam	X Consultants
Mechanical engineer Bridget James	RG Consultants
Electrical engineer Walter Ampere	PS Design
Soil mechanic engineer	To be appointed during October 2011

1.10   Communication management

Project server and approvals for project members' accessibility	PM
Reports to project owner and core stakeholders	Every fifth week
Archives	Lind Rohn

1.11   Risk management

Risk analyses shall be done for:
The whole project and be included in the investment proposal for design
The whole project and be included in the investment proposal for production
Design work
Production (together with contractor)
Start-up of standby power units

1.12   Procurement management

DBC contract
Elevators purchased early and transferred to main contractor
Standby power units purchased early and transferred to main contractor

1.13   Approval management

FAT for standby power units
SAT1 and SAT2 for all main subcontracts

APPENDIX B:
WBS TEMPLATE

Scope description

Activity description	
Input	Assumptions:
Dependent on other activities	
Process/method	
Quality activities	
Contract references	Needed approvals:

Time management tider

	Earliest acceptable:	Latest acceptable finish
Activity duration		
Internal milestones		Date
External milestones		Date

Resource management resursuppskin

Resources	Type	Period	Number	Cost per hour	Total cost
Personal cat.1					
Personal cat.2					
Material					
Other					

Risk

Risks	

APPENDIX C:
PIT INFORMATION

Building	Room	Machinery, equipment, etc.	Clear opening			Slope	Facing layer	Floor drain material	Incoming/outgoing water, heat, other	Incoming/outgoing electricity
			Length	Width	Depth					

APPENDIX D:
ELEVATOR INFORMATION

	Project name/no.:					Date:		
						Responsible:		
Elevator no.	Type*	Cable, hydraulic, other	Lifting height	Load	Speed	Int./ ext. phone	Type of door	

* P = Passenger lift; GP = Goods/passenger lift; G = Goods lift; L = Lift table.

Elevator no.	Part	Width	Depth	Height (up/down)	Other
	Elevator car				
	Elevator well				
	Well top				
	Well bottom				
	Machine room				
	Clear door opening				
	Outside frame dimension				
	Free area on side for two-leaf side opening doors				
	Clear door opening for machine room				

Elevator no.	Issue	Comment
	Rest. floor when fire	
	Spec. door function	
	Fault signal to:	
	Alarm signal to:	

APPENDIX E:
DOOR INFORMATION

	Project name/no.:	Date:	
		Responsible:	
Door type (description/spec.)			
Number			
Ref. to drawing			
Function. Wood, steel, glazed, double, evacuation, fire, mechanical door; right- or left-hung door; self-closing			
Type			
Material, door leaf			
Material, door frame			
Material, door sill			
Frame dimensions			
Fire rating			
Sound-insulation class			
Glass openings			
Surface finish			
Hinge; flat, rising…			
Door closer (think of door size and weight)			
Location of door opener (handicap code?)			
Panic device			
Door handle			
Flush bolt			
Location of security opening device			
Electrical conduits in frame and door leafs			
Lock case			
Lock			
Lock cylinder (who purchases, installs?)			
Lock accessories			
Others			

APPENDIX F:
INFORMATION EXCHANGE INTERFACES

Project name/no:

Date:

Responsible:

Equipment/Machinery	Specification information not later than	Responsible for requirement specification	Information purchaser		Architect	Structural design engineer	Mechanical/ design engineer	HVAC design engineer	Electrical design engineer
			Received	Given					
Preliminary									
Size, weight									
Heat loads									
Pits, drains, lift table									
Floor services, electrical, water, etc.									
Effects, steam pressure/flow, and cooling									
Air flow rate									
Particle filtering device									
Location of service connection									
Final information									
Size, weight									
Heat loads									
Pits, drains, lift table									
Floor services, electrical, water, etc.									
Effects, steam pressure/flow, and cooling									
Air flow rate									
Particle filtering device									
Location of service connection									

APPENDIX G:
INTERFACE INFORMATION FOR EQUIPMENT AND MACHINERY

	Project name/no.:		Date:	
			Responsible:	
	Machinery/ equipment	Service designed by	Responsible for purchase of work	Work done by
Construction work				
Electrical installations				
Ventilation				
Cold water				
Hot water				
Steam				
Sewage				
Brine				
Compressed air				
Others				

Note: Services from building, electrical, water, gas, ventilation etc., contractors.

APPENDIX H:
CONTRACT LIMITS (1/3)

	Responsible for requirement specification	Responsible designer	Design coordination with	Responsible for coordination of design	Part of contract no.	Delivered by	Installed by	Connected by	Comment
Terrain									
External sewage									
External water installations									
External electrical installations									
Ground work under road surfacing and garden work									
Construction									
Decontamination									
Demolition									
Close and insulate openings after demolition									

APPENDIX H:
CONTRACT LIMITS (CONTINUED 2/3)

	Responsible for requirement specification	Responsible designer	Design coordination with	Responsible for coordination of design	Part of contract no.	Delivered by	Installed by	Connected by	Comment
Dust sealing									
Protection of prefab stairs during production									
Adjustment of fire alarm									
Fan/pump foundation									
External signs									
Mounting device for external signs									
Signs outside rooms									
Archive fitments									
Fire protection blankets									
Portable fire extinguishers									

(continued)

APPENDIX H:
CONTRACT LIMITS (CONTINUED 3/3)

	Responsible for requirement specification	Responsible designer	Design coordination with	Responsible for coordination of design	Part of contract no.	Delivered by	Installed by	Connected by	Comment
Mechanical/HVAC									
Sprinkler									
Electrical									
Standby power									
External lighting									
IT installations									
ITV installation									
Others									
Traverse crane									
Lifting crane									
Lifting table									
Compacting machine									

APPENDIX I:
CHANGE AGREEMENT

	Project name/no.:	Date:	
		Responsible:	
Client:			
Contractor:			

	Order		
The following has been ordered:	Memo no.:	Drawing/sketch	
Remuneration		Unit price/rate	
		Fixed price as offered	
		Cost plus with price ceiling	Price:
		Cost plus	
		Prices will not be indexed	
Client		Contractor	

Cost specification			Page:
Date	Work	Memo, drawing, etc.	$
		Total/transport	

Price above is agreed			
Place:	Date:	Place:	Date:
Client		Contractor	

APPENDIX J:
CHANGE LOG

	Project name/no.:			Date:			
				Responsible:			
	Estimated design cost		**Estimated production cost**				
Change description	Designer	Cost	Contr. no.	Cost	Approved at meeting no.	Rejected at meeting no.	

APPENDIX K:
RISK MANAGEMENT PLAN

Risk Manager:
Participants:

Impact:

1 = Very low
2 = Low
3 = Medium
4 = High
5 = Very high

Probability:

1 = Probably not occur
2 = Low probability
3 = Can appear
4 = Likely
5 = Most likely

Event	Impact	Before RR			Risk response	After RR			Date	Resp.
		I	P	Rx		I	P	Rx		

Index